Physical Principles of Remote Sensing

TOPICS IN REMOTE SENSING

Series editors
Garry Hunt and Michael Rycroft

Physical Principles
of Remote Sensing

W. G. REES

Scott Polar Research Institute
University of Cambridge

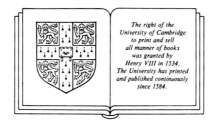

The right of the
University of Cambridge
to print and sell
all manner of books
was granted by
Henry VIII in 1534.
The University has printed
and published continuously
since 1584.

CAMBRIDGE UNIVERSITY PRESS

Cambridge

New York Port Chester Melbourne Sydney

Published by the Press Syndicate of the University of Cambridge
The Pitt Building, Trumpington Street, Cambridge CB2 1RP
40 West 20th Street, New York, NY 10011, USA
10 Stamford Road, Oakleigh, Melbourne 3166, Australia

© Cambridge University Press 1990

First published 1990

Printed in Great Britain at the University Press, Cambridge

British Library Cataloguing in publication data
Rees, W.G.
Physical principles of remote sensing
1. Remote sensing
I. Title II. Series
621.367

Library of Congress cataloguing in publication data
Rees, W.G.
Physical principles of remote sensing / W.G. Rees.
p.   cm. – (Topics in remote sensing)
Includes bibliographical references.
ISBN 0 521 35213 4. – ISBN 0 521 35994 5 (pbk.)
1. Remote sensing.   I. Title.   II. Series.
G70.4.R44   1990
621.36'78–dc20   90–1455 CIP

ISBN 0 521 35213 4 hardback
ISBN 0 521 35994 5 paperback

VN

For Christine

CONTENTS

PREFACE

There are many books which explain the subject of Remote Sensing to those whose backgrounds are primarily in the environmental sciences. This is an entirely reasonable approach, since these are the main users of remotely sensed data. However, as the subject grows in importance the need for a significant number of people to understand not only what remote sensing systems do, but how they work, will grow with it. This is already happening as increasing numbers of physical scientists, engineers and mathematicians move into the field of environmental remote sensing, and it is for such readers that this book has been written. That is to say, the reader for whom I have imagined myself to be writing is educated to a reasonable standard (although not necessarily to first degree level) in physics, with a commensurate level of mathematical ability. I have however found it impossible to be strict about this, because of the wide range of disciplines within and beyond physics from which the material has been drawn, and I trust that users of this book will be duly indulgent when they find it either too simple, or over their heads.

This book attempts to follow a logical progression, more or less reflecting the flow of information from the remotely sensed object to the user of the data. The first two chapters lay the general foundations of the subject. Chapter 2 is a non-rigorous treatment of electromagnetic theory, which can be regarded as a compendium of necessary results. It will represent, I hope, mostly revision to most readers, although it assumes little or no knowledge of Fourier transforms or Fraunhofer diffraction theory. Chapter 3 discusses the interaction of electromagnetic radiation with rough surfaces and with the atmosphere, so that now our information is heading upwards towards the sensor, as it were.

Chapters 4 to 8 discuss the sensors themselves, 4 to 6 being devoted to passive systems and 7 and 8 to active systems. These chapters explain, so far as is consistent with the level of the book, the functioning of the sensors, important operational constraints, and some of the more important applications of the data derived from them.

The platforms on which the sensors are supported are discussed in chapter 9. After a short discussion of remote sensing from aircraft, the chapter devotes itself to satellite orbits. Finally chapter 10 presents an introduction to the data processing involved in remote sensing, particularly image processing and analysis. An appendix contains tables of data needed frequently in remote sensing. A short list of problems is included at the end of each chapter, and answers to the numerical problems are provided at the back of the book. Most of the problems are straightforward, and designed to consolidate and extend the reader's understanding of the material of the chapter. Some problems require material from more than one chapter.

Some notes on the rationale of this book are in order. It has been my intention to keep the book as short as possible, consistent with clarity. In particular, since the book is intended to teach the *principles* of remote sensing, it avoids presenting too much technical or 'engineering' detail. For example, the LANDSAT MSS scanner system is described in chapter 5 in a schematic way only. The same principle has been applied to the provision of the bibliography, which is intended to provide a comprehensive *entrée* into the literature of remote sensing, but not to overwhelm the reader with an enormous list of references. Some selection and omission has therefore been necessary, and I hope that my scientific colleagues will forgive me if my selection does not tally with theirs. So far as the use of symbols in equations is concerned, the book deliberately avoids the rigorous consistency which demands that a given symbol be used to represent only one physical quantity. Because of the wide scope of remote sensing, this would lead to an unforgivably confusing proliferation of symbols with many sub- and superscripts. Consistency of symbols is, therefore, confined to sections of the text which deal with a single topic, except for a few 'universal' symbols (such as h for Planck's constant and ω for angular frequency) which are used throughout the book. The S.I. system is used consistently, although a table in the appendix gives equivalents for some common non-S.I. units.

This book arose from a course of undergraduate lectures delivered first at the Scott Polar Research Institute and then at the Cavendish

Laboratory, both in the University of Cambridge. I am grateful to the heads of both departments for letting me try out my ideas. I am grateful also to Dr Caroline Roberts of Cambridge University Press, and Dr Michael Rycroft, series editor of the Topics in Remote Sensing, for advice and encouragement. Many other individuals are owed thanks for their contributions, witting or otherwise, to the writing of this book. Those whom I particularly wish to thank, however, are Dr Andrew Cliff, Dr Bernard Devereux, Michael Gorman and Christine Rees. Each will, I hope, understand his or her part in this work. I naturally take responsibility for the errors, obscurities and infelicities which remain, and would be grateful to have them pointed out to me.

W.G.R.
Cambridge

ACKNOWLEDGEMENTS

Permission to reproduce copyright and other material from the following sources is gratefully acknowledged: Cambridge University Committee for Aerial Photography (United Kingdom), Centre National des Etudes Spatiales (France), Deutsche Forschungs- und Versuchsanstalt für Luft- und Raumfahrt, European Space Agency, National Aeronautics and Space Administration (USA), National Remote Sensing Centre (United Kingdom), Nigel Press Associates Ltd (United Kingdom), Norwegian Defence Research Establishment, Plenum Press (USA), University of Dundee, Dr T. Wahl.

1

Introduction

1.1 Definitions

Remote sensing is, broadly but logically speaking, the collection of information about an object without coming into physical contact with it. This definition is too wide to be useful, so we shall impose a number of restrictions. The first is that the object is located on or near the earth's surface, and the sensor is more or less above the object, and at a 'substantial' distance from it. The second restriction is that the information is carried by electromagnetic radiation, some property of which is affected by the remotely sensed object.

Already we have narrowed the scope of our definition so as to exclude such techniques as sonar, seismic and geomagnetic measurements, medical imaging, and the remote measurement of the core temperatures inside nuclear reactors, all of which are commonly referred to by their practitioners as remote sensing techniques, and yet the definition is already vague and unwieldy. In fact, remote sensing is a subject which is notoriously difficult to define in a satisfactory manner (e.g. Cracknell, 1981). We shall not, therefore, worry unduly about the lack of rigour of our definition, and content ourselves with the operational definition that what is contained in this book is what we mean by remote sensing.

One final remark should be made, though, before we leave this section. Remote sensing, even in the restricted sense in which we have tried to define it, usually includes the study of the earth's atmosphere as well as its surface. For reasons of space we shall not discuss atmospheric sounding, except briefly.

1.2 The need for remote sensing

The principal advantages of remote sensing are the speed at which data can be acquired from large areas of the earth's surface, and the related fact that comparatively inaccessible areas may be investigated in this way. The uses, both existing and potential, of such data within the various environmental disciplines are legion. A good summary of the applications of remote sensing to environmental problems is given by Barrett & Curtis (1982). It is impossible to list all the applications, but is perhaps revealing to mention a few (in no particular order):

Meteorology, e.g.
> profiling of atmospheric temperature, pressure and water vapour content, measurement of wind velocity.

Oceanography, e.g.
> measurement of the sea's surface temperature, mapping ocean currents and wave energy spectra.

Glaciology, e.g.
> mapping the distribution and motion of ice sheets and sea ice, determining the navigability of sea ice.

Geology, geomorphology and geodesy, e.g.
> identification of rock type, location of geological faults and anomalies, measuring the figure of the earth and observing tectonic motion.

Topography and cartography, e.g.
> obtaining accurate elevation data and referring them to a given coordinate system, production and revision of maps.

Agriculture, forestry and botany, e.g.
> monitoring the extent and type of vegetation cover and its state of health, identifying the host plants of pests, mapping soil type and determining its water content, forecasting crop yields.

Hydrology, e.g.
> assessing water resources, forecasting meltwater run-off from snow.

Disaster control, e.g.
> warning of sand and dust storms, avalanches, landslides, flooding etc., monitoring the extent of floodwater, monitoring of pollution.

Planning applications, e.g.
> generation of inventories of land use and monitoring changes, assessing resources, performing traffic surveys.

Military applications, e.g.

    monitoring the movement of vehicles and military formations, assessing terrain.

Remote sensing, especially when conducted from space, is an intrinsically expensive activity. Nevertheless, cost-benefit analyses demonstrate its financial effectiveness, and much speculative or developmental remote sensing activity can be justified in this way. The true cost (i.e. without subsidy) of developing a remote sensing satellite and putting it into a low earth orbit is probably of the order of £10^8. Although the potential benefits are harder to quantify, figures totalling about £5×10^7 have been estimated (Erich & Gottschalk, 1984) for the marine environment alone in the case of the ERS-1 satellite, whose mission is scheduled to last only two years, and crop inventory data derived from satellite remote sensing can be worth of the order of £10^8 per annum to a country which invests in this market (Slater, 1980). Colwell (1983) gives a similar analysis for the use of aerial photography in assessing forest resources, which suggests that aerial surveying capable of yielding a planimetric map at a scale of 1 : 50 000 costs about one tenth of the probable exploitation value of the data. The use of satellite remote sensing data, particularly from meteorological satellites, for disaster warning has already saved thousands of human lives.

1.3 Historical sketch of remote sensing

Remote sensing can plausibly be traced back to the fourth century BC and Aristotle's camera obscura (or, at least, the instrument described by Aristotle in his *Problems*, but perhaps known earlier). Although significant developments in optical theory began to be made in the seventeenth century, and glass lenses were known much earlier than this, practical remote sensing had to await the invention of photography (by Fox Talbot and Daguerre, following the earlier work of Wedgwood) in the first half of the nineteenth century. During the remarkable nineteenth century, also, forms of electromagnetic radiation beyond the visible were discovered by Herschel (infrared), Ritter (ultraviolet) and Hertz (radio waves), and in 1863 Maxwell developed the electromagnetic theory upon which so much of our understanding of these phenomena depends.

Airborne photography followed almost immediately on the discovery of the photographic method. In 1858, Gaspard Felix Tournachon ('Nadar') made what was probably the first aerial photograph, unfortunately no longer in existence, from a balloon at an altitude of about 80 m.

Kites were also soon used, and by 1890 the usefulness of aerial photography was so far recognised that Batut had published a textbook on the subject.

The next step in the development of the subject was taken with the invention of the aeroplane in 1903. Again, the potential applications were quickly seen and aerial photographs from aeroplanes were recorded from 1909. Airborne photography was used during the First World War for military reconnaissance, thus establishing the fairly close connexion between environmental and military remote sensing which has existed since then. During the period between the two World Wars, civilian uses of aerial photography began to develop, notably in cartography, geology, agriculture and forestry. Cameras improved, as did aircraft, and stereographic mapping attained an advanced state of development. Also during this period, early work was performed by John Logie Baird, the inventor of television, on the development of airborne scanning systems capable of transmitting images to the ground. This work was highly confidential, having been carried out on behalf of the French Air Ministry. It was ended by the war and forgotten about until 1985 (Newton, 1989).

The Second World War brought substantial developments to remote sensing. Photographic reconnaissance reached a high state of development. The German invasion of Britain planned for September 1940 was forestalled by the observation of concentrations of ships along the English Channel. At the same time, infrared-sensitive instruments and radar systems were being developed. In particular, the Plan Position Indicator (PPI) used by night bombers was an imaging radar which presented the operator with a 'map' of the terrain, and thus represents the ancestor of the side-looking airborne radar and synthetic aperture radar systems discussed in chapter 8.

By the 1950s, false-colour infrared film, developed for military purposes, was finding applications in vegetation mapping, and high-resolution imaging radars were being developed. As these developments continued through the 1960s, sensors began to be placed in space. This was originally part of the programme to observe the moon, but the possibility of applying the same technique to observation of the earth was soon realised, the first multispectral spaceborne imagery being obtained from Apollo 6. Since that time, radars have been placed in space (this step had to await technological improvements in radar design in order to provide systems with low mass and high spatial and radiometric resolution), and the computer revolution has made digital image processing a real possibility. This may be regarded as the beginning of the 'modern era' of remote sensing, in which airborne and spaceborne observations have become routine.

1.4 Classification of remote sensing systems

Remote sensing systems may be classified in a number of ways, but the most useful distinctions we can draw are between *active* and *passive* systems, and between *imaging* and *non-imaging* systems. We may also distinguish sensors by the wavelength of the radiation to which they respond.

Active systems illuminate the object of study with their own supplied radiation, whereas passive systems sense naturally occurring (emitted thermal or reflected solar) radiation. The choice between active and passive systems is influenced by a number of factors. A passive system will be inappropriate at certain wavelengths at which insignificant amounts of radiation occur naturally. An active system may be technically infeasible if the amount of power which has to be radiated in order to obtain a measurable reflected signal is too great. It may be desirable to have exact knowledge of the nature of the illuminating radiation, which can be tailored to some particular aspect which is to be studied. (For example, we may wish to observe Doppler shifts in the reflected radiation, so as to be able to calculate the relative motion of the target and the sensor.)

An imaging system is slightly harder to define. We shall take the term to mean a system, either active or passive, which measures the intensity of the radiation reaching it and which does so as a function of position on the earth's surface so that a two-dimensional pictorial representation of the intensity can be constructed. A non-imaging system is thus one which either does not measure radiation intensity, or does not do so as a function of position on the earth's surface. We are thus employing a rather restricted definition of the word 'image'. Note that the spatial condition on an imaging system can be reduced to a statement that, for a single location of the system, it should measure the radiation intensity from a number of discrete regions distributed in one or two dimensions. One dimension is adequate for the production of an image, since motion of the platform (on which the system is supported) in the perpendicular direction achieves the necessary two-dimensional scanning (see fig. 1.1).

Fig. 1.2 illustrates the division of remote sensing systems according to whether they are active or passive, imaging or non-imaging. The figure also attempts to show that the definition provided in the previous paragraph is somewhat fuzzy. Nevertheless, the division is a useful one which will influence the structure of the book.

Problem

1. Explain what is meant by 'remote sensing'. What advantages does it offer?

Fig. 1.1. Imaging systems measure the intensity of the radiation from each cell (resolution element) on the earth's surface. As the field of view is carried forward by the platform, a two-dimensional image is built up.

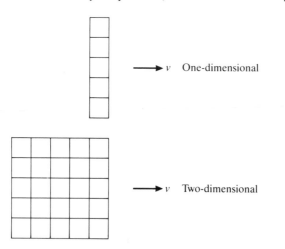

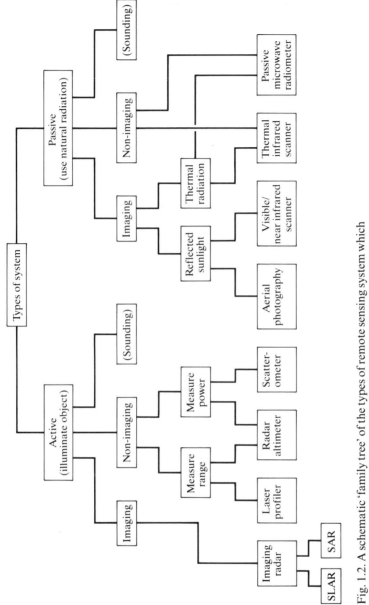

Fig. 1.2. A schematic 'family tree' of the types of remote sensing system which are considered in this book. Sounding systems, which yield profiles of the atmosphere, are not discussed in detail.

2

Electromagnetic radiation in homogeneous media

2.1 Introduction

In this chapter we discuss the basic principles of electromagnetic propagation, and of the thermal generation of radiation. The treatment is not rigorous, in that not all of the results are derived, and indeed much of the chapter may be regarded as a review of the necessary material. Many excellent textbooks exist which supplement this approach. Particularly useful are Bleaney & Bleaney (1976), Hecht (1987) and, at a somewhat deeper level, Jackson (1975). The book by Lipson & Lipson (1981) also contains many illuminating insights into optical physics.

The one notable exception to the statement that most of the material in this chapter is presented in review form is the treatment of frequency spectra and diffraction. The theory of Fourier transforms is sufficiently vital to the subject of image science, and may be unfamiliar to some readers, so that it seems appropriate to include a reasonably complete derivation here. Further information on this topic can be found in the book by Bracewell (1978).

This chapter is unusual in that the level of mathematical competence assumed is somewhat greater than in most other parts of the book. It is hoped that the less mathematically inclined reader will persevere with it (or skim it selectively) in order to obtain a rough idea of its material and so make sense of subsequent chapters.

2.2 Basics

The propagation of electromagnetic radiation as waves is a consequence of the form of Maxwell's equations, as Maxwell himself realised.

(Longair, 1984 provides a very readable account of the development of Maxwell's electromagnetic theory.) One form in which these equations may be written is, in the absence of free charges,

$$\left.\begin{array}{l} \mathbf{V} \cdot \mathbf{E} = 0 \\ \mathbf{V} \cdot \mathbf{B} = 0 \\ \mathbf{V} \wedge \mathbf{E} + \partial \mathbf{B}/\partial t = 0 \\ \mathbf{V} \wedge \mathbf{B} - \mu\varepsilon/c^2 \, \partial \mathbf{E}/\partial t = 0 \end{array}\right\} \tag{2.1}$$

In these expressions, $\mathbf{E}$ is the wave electric field and $\mathbf{B}$ the wave magnetic field, and $c = (\varepsilon_0\mu_0)^{-\frac{1}{2}}$ where ε_0 and μ_0 are the *electric permittivity* and *magnetic permeability* of free space. μ and ε are respectively the relative magnetic permeability and the relative electric permittivity of the medium, and both are dimensionless. The relative electric permittivity is also known as the *dielectric constant*. It can easily be shown (using the vector operator identity $\mathbf{V} \wedge (\mathbf{V} \wedge \mathbf{A}) = \mathbf{V}(\mathbf{V} \cdot \mathbf{A}) - \mathbf{V}^2\mathbf{A}$) that $\mathbf{E}$ and $\mathbf{B}$ must satisfy the equations

$$\mathbf{V}^2\mathbf{E} - \mu\varepsilon/c^2 \, \partial^2\mathbf{E}/\partial t^2 = 0 \tag{2.2}$$

$$\mathbf{V}^2\mathbf{B} - \mu\varepsilon/c^2 \, \partial^2\mathbf{B}/\partial t^2 = 0 \tag{2.3}$$

These are three-dimensional wave equations. It is clear, because of the similarity of these equations, that we will not sacrifice any generality by dealing only with the electric field $\mathbf{E}$. We can easily show that a harmonic wave satisfies (2.2), although we need to use again the fact that $\mathbf{V} \cdot \mathbf{E}$ is zero to establish that the vector $\mathbf{E}$ must always be perpendicular to the direction of propagation. Without loss of generality, then, we can try as a solution of (2.2) a wave of the form

$$E_x = E_0 \cos(\omega t - kz) \tag{2.4}$$

where we have chosen the propagation to be along the z-axis and the orientation of the $\mathbf{E}$ vector to be along the x-axis. (We have used here the *angular frequency* ω and the *wavenumber* k, rather than the more familiar cyclic frequency v or f and wavelength λ. The former are usually more useful, and we shall use them often.) Substituting (2.4) into (2.2) gives

$$\omega = ck(\varepsilon\mu)^{-\frac{1}{2}} \tag{2.5}$$

or, since $\omega = 2\pi v$ and $k = 2\pi/\lambda$,

$$v = c\lambda^{-1}(\varepsilon\mu)^{-\frac{1}{2}}$$

Thus we see that the form of E_x in (2.4) satisfies the wave equation (2.2), and that the wave velocity is

$$v = c(\varepsilon\mu)^{-\frac{1}{2}} \tag{2.6}$$

where $c = (\varepsilon_0 \mu_0)^{-\frac{1}{2}}$. c is the speed of light *in vacuo*, and has a value of $2.99792458 \times 10^8 \text{ ms}^{-1}$. (This value is very well determined, and in fact now defines the metre in terms of the second. Values of important constants such as c are given in the appendix.) Equation (2.6) is often written as

$$v = c/n \qquad (2.7)$$

where n is termed the *refractive index*, and clearly $n = (\varepsilon \mu)^{\frac{1}{2}}$.

So far we have discussed the propagation of an electromagnetic wave only in terms of the electric field vector **E**. However, exactly the same reasoning applies to the magnetic field **B**. A simple harmonic wave of the form

$$B = B_0 \cos(\omega t - kz)$$

describes its variation, and, like the **E** vector, the **B** vector is perpendicular to the direction of propagation. In fact, the **E** and **B** fields can also be shown to be perpendicular to each other, and the directions to be such that the vector $\mathbf{E} \wedge \mathbf{B}$ is parallel to the propagation direction. Thus for example if the propagation direction is along z, and the electric field is along x, then the magnetic field must be along y in a right-handed system of axes.

E_0 and B_0 are the *amplitudes* of the electric and magnetic fields of the wave. Since the **E** and **B** fields oscillate in phase,

$$E/B = E_0/B_0 = v = Z/\mu\mu_0 \qquad (2.8)$$

where Z is a constant for a given medium. Z is called the *impedance*, and has a value of

$$Z = (\mu\mu_0/\varepsilon\varepsilon_0)^{\frac{1}{2}} = Z_0(\mu/\varepsilon)^{\frac{1}{2}} \qquad (2.9)$$

where Z_0 is called the *impedance of free space*. Clearly it is given by $Z_0 = (\mu_0/\varepsilon_0)^{\frac{1}{2}}$, and has a value of approximately $377\,\Omega$. The mean energy *flux density* (i.e. power crossing unit area normal to the direction of propagation) of the electromagnetic wave is given by

$$F = E_0^2/2Z, \qquad (2.10)$$

and it is conventional to refer to E_0 as the amplitude of the wave.

We have not yet said anything about the frequency of the radiation which we are discussing. What distinguishes one type of radiation from another is its frequency or, equivalently, its wavelength, and the whole range of frequencies is called the *electromagnetic spectrum*. Different regions of the spectrum are given different names such as light, radio waves, gamma radiation and so on, usually referring to the manner in which the radiation is detected or generated, but these names are of course purely conventional. The electromagnetic spectrum is shown schematically in fig. 2.1.

Fig. 2.1. The electromagnetic spectrum. The diagram shows those parts of the electromagnetic spectrum which are important in remote sensing, together with the conventional names of the various regions of the spectrum. The letters (*P, L, S* etc.) used to denote parts of the microwave spectrum are in common use in remote sensing, being standard nomenclature amongst radar engineers in the USA. Note that this nomenclature varies somewhat in other countries, particularly in military usage. Note also that various terminologies are in use for the subdivisions of the infrared (IR) part of the spectrum. That adopted here defines the thermal infrared band as lying between 3 and 15 μm, since this region contains most of the power emitted by black bodies at terrestrial temperatures.

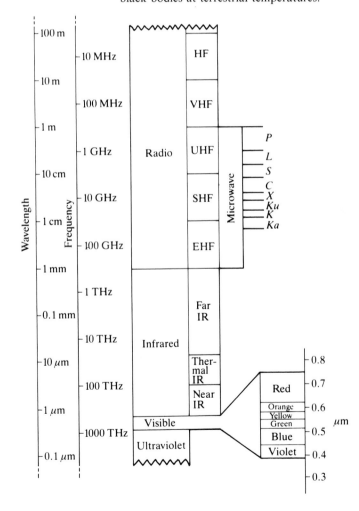

We remarked earlier that the speed of electromagnetic waves is given by $v = c/n$ (2.7), where the refractive index $n = (\varepsilon\mu)^{\frac{1}{2}}$. In all the media we shall consider, $\mu = 1$ (i.e. the media are non-magnetic), and we can put $n = \varepsilon^{\frac{1}{2}}$. This is straightforward if ε is a real positive number, and this describes a medium in which the electromagnetic wave is not absorbed. However, if the medium is *lossy* and the wave is absorbed by it we need to use a complex number for the relative permittivity. The conventional way of doing this is to put

$$\varepsilon = \varepsilon' - i\varepsilon'' \tag{2.11}$$

(where $i^2 = -1$), which is equivalent to making the refractive index complex:

$$n = n' - i\kappa \tag{2.12}$$

(We have used the Greek letter kappa to represent the negative of the imaginary part of the refractive index.) Note that (2.11) is also sometimes written

$$\varepsilon = \varepsilon'(1 - i\tan\delta), \tag{2.13}$$

where $\tan\delta$ is the *loss tangent*. Simple *non-polar* materials have constant values of ε' and ε'', and simple *polar* materials are described by the *Debye equation* (2.14), which represents a resonant phenomenon with a time-constant τ. Fig. 2.2 illustrates this type of behaviour for water, which obeys the Debye equation fairly closely between about 10^6 and 10^{12} Hz.

$$\left.\begin{array}{l} \varepsilon' = \varepsilon_\infty + \varepsilon_p/(1 + \omega^2\tau^2) \\ \varepsilon'' = \omega\tau\varepsilon_p/(1 + \omega^2\tau^2) \end{array}\right\} \tag{2.14}$$

In these expressions, ε_∞, ε_p and τ are constants.

The two other types of dielectric behaviour which we need to identify are those of conducting media and of plasmas. In a simple *conductive* medium, for which we assume that polar contributions to the dielectric properties are negligible, the imaginary part of the dielectric constant is related to the conductivity σ through

$$\varepsilon'' = \sigma/\varepsilon_0\omega \tag{2.15}$$

In a *plasma*, in which all atoms have been ionized, the dielectric constant is given by

$$\varepsilon = n^2 = 1 - Ne^2/\varepsilon_0 m\omega^2 \tag{2.16}$$

In this expression, N is the number density (i.e. measured in units of length^{-3}), e is the charge and m is the mass of the particles of the plasma.

Because the mass of the electron is so much smaller than that of any other charged particle, the latter may effectively be ignored in applying this expression. It is clear from this expression that the refractive index of a plasma is purely real at high frequencies, and purely imaginary at low frequencies. This fact will be of importance when we consider the effects of the ionosphere.

The types of dielectric behaviour which we have been discussing are, in general, idealisations. Real materials can often be understood by invoking different models over different ranges of frequency.

It is easy to show from (2.11) and (2.12), using the fact that $\varepsilon = n^2$, that

$$
\left.
\begin{aligned}
&\varepsilon' = n'^2 - \kappa^2 \\
&\varepsilon'' = 2n'\kappa \\
&n'^2 = (\varepsilon' + [\varepsilon'^2 + \varepsilon''^2]^{\frac{1}{2}})/2 \\
&\kappa = \varepsilon''/2n'
\end{aligned}
\right\}
\qquad (2.17)
$$

These relationships are illustrated by fig. 2.3.

We can see that a complex refractive index (2.12) describes a wave which is absorbed by the medium, by using the complex exponential form of (2.4) to represent the wave. If we put

$$
\begin{aligned}
E &= E_0 \exp\{i(\omega t - kz)\} \\
&= E_0 \exp\{i\omega(t - nz/c)\}
\end{aligned}
$$

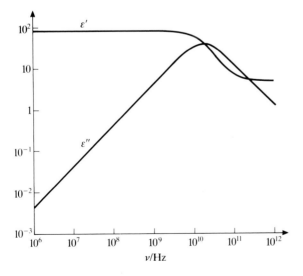

Fig. 2.2. The real and imaginary parts of the dielectric constant of pure water at 20 °C.

(which is equivalent to (2.4) if we observe the convention that the *real part* of the complex number represents the corresponding physical quantity) and substitute (2.12) for n, we find that

$$E = E_0 \exp\{i\omega(t - n'z/c)\}\exp(-\omega\kappa z/c) \tag{2.18}$$

Equation (2.18) clearly represents a simple harmonic wave whose amplitude decreases exponentially with distance z. The *absorption length* l_a is usually defined as the distance travelled by the wave over which its amplitude falls by a factor of e (and, since by (2.10) the intensity is proportional to the square of the amplitude, the intensity falls by a factor of e^2 which is about 7.4), so this is given by

$$l_a = c/\omega\kappa = \lambda_0/2\pi\kappa \tag{2.19}$$

where λ_0 is the free-space wavelength of the radiation.

Many materials, although lossy, are not exceptionally so and we can put $\varepsilon'' \ll \varepsilon'$ (or, equivalently, $\kappa \ll n'$ or $\delta \ll 1$). In this case, (2.17) can be simplified to

$$\left.\begin{array}{l} n' \approx \sqrt{\varepsilon'} \\ \kappa \approx \varepsilon''/2\sqrt{\varepsilon'} \end{array}\right\} \tag{2.20}$$

so

$$l_a \approx \lambda_0\sqrt{\varepsilon'}/\pi\varepsilon'' \tag{2.21}$$

Fig. 2.3. The relationship between dielectric constant ($\varepsilon' - i\varepsilon''$) and refractive index ($n' - i\kappa$).

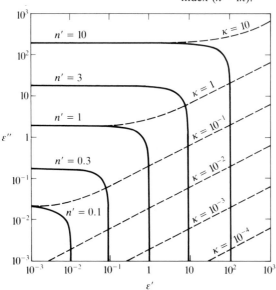

2.2.1 Dispersive waves

We noted earlier that, in a number of cases of practical importance, the dielectric properties of a medium, and hence its refractive index, vary with frequency. Such media are said to be *dispersive*, and a wave propagating in such a medium is called a *dispersive wave*. It is usual to characterise this behaviour by expressing the angular frequency ω as a function of the wavenumber k, and this relationship is called the *dispersion relation*.

We saw in (2.5) and (2.6) that the wave velocity v is given by

$$v = \omega/k \tag{2.22}$$

This is the speed at which the crests and troughs of a continuous wave travel in the direction of propagation. If we *modulate* the wave in some way, for example by breaking it up into pulses, it is the modulation which carries information, and we therefore need to know the speed at which this modulation travels. This is the *group velocity*, and it is given by

$$v_g = d\omega/dk \tag{2.23}$$

Only in the case of a non-dispersive wave, for which ω is proportional to k, will the two velocities be in general equal to one another.

It will often happen in practice that the dispersion relation will be expressed not as $\omega(k)$, but as $n(\lambda_0)$, where n is the refractive index and λ_0 is the free-space wavelength (i.e. $\lambda_0 = 2\pi c/\omega$). In this case, (2.23) may be expressed as

$$c/v_g = n - \lambda_0 dn/d\lambda_0 \tag{2.24}$$

It can be shown from this expression that if the refractive index satisfies an equation of the form $n^2 = 1 - A\lambda_0^2$ (A being a constant), which as we have seen is true for plasmas (2.16), the group and wave velocities are related by

$$vv_g = c^2 \tag{2.25}$$

Since the refractive index of a plasma is less than unity, the wave velocity is greater than the speed of light. Equation (2.25) shows, however, that the group velocity must therefore be less than the speed of light. This is of course in conformity with Einstein's relativistic postulate.

2.2.2 Polarisation

We noted earlier that the E and B fields in an electromagnetic wave are perpendicular to each other, and to the direction of propagation. This does not however uniquely define their orientation. The orientation of the fields is termed the *polarisation* of the radiation, and, as we shall see, it is important to consider it in discussing the operation of a remote sensing system.

If the **E** vector remains in the same plane (and the **B** vector therefore remains in its own perpendicular plane), the radiation is said to be *plane polarised*. This is illustrated in fig. 2.4. Although in principle the orientation could be specified using either **E** or **B**, it is conventional to use the electric field, so the example of fig. 2.4 would be described as vertically (or *x*) polarised. If the **E** and **B** vectors instead rotate about the propagation axis such that the end of each vector describes a helix (fig. 2.5), the radiation is called *circularly polarised*. Such polarised radiation can be either left-hand circularly polarised (LHC) or right-hand circularly polarised (RHC), depending on whether the helix is anticlockwise or clockwise when viewed along the direction of propagation. The only other kind of 'pure' polarisation is *elliptically polarised* radiation, in which the **E** and **B** vectors describe elliptical helices. In general, the polarisation of an electromagnetic wave will be a mixture of these various types, and may include a *randomly polarised* component in which the orientation of the **E** vector changes randomly. This kind of radiation is often called *unpolarised* radiation, although this is a somewhat misleading term since it suggests that the **E** vector does not point in any direction.

Fig. 2.4. Plane polarised radiation. The wave is propagating in the *z*-direction and is polarised with the electric field parallel to the *x*-axis and the magnetic field parallel to the *y*-axis. The straight lines represent the instantaneous magnitudes and directions of the fields, although the amplitudes E_0 and B_0 have been given different scales for clarity.

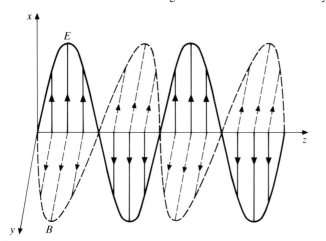

Most natural sources of radiation are randomly polarised, although as we shall see scattering or reflexion may change the state of polarisation. If randomly polarised radiation is incident on a device which detects only one polarisation (such as LHC and not RHC, or horizontally polarised and not vertically polarised), only half of the available power will be collected. If the detector responds only to (say) LHC polarised radiation and the incident field is entirely RHC, nothing at all will be detected. It is thus important to consider polarisation in the design of any remote sensing system.

2.2.3 *The Doppler effect*

If a source of electromagnetic radiation of frequency v is in motion with respect to an observer, the observer will detect the radiation at a different frequency v'. If the source is approaching the observer, or equivalently if the observer is approaching the source, v' will be greater than v, and conversely. This is known as the Doppler effect, and is analogous to the similar (and familiar) effect observed with sound waves. However, whereas the Doppler effect for sound is not the same for the source approaching the observer and for the observer approaching the source,

Fig. 2.5. Right-hand circularly polarised radiation. The notation is the same as in fig. 2.4, although the magnetic field vectors have been omitted for clarity. They are, as always, oriented perpendicularly to the electric field vectors.

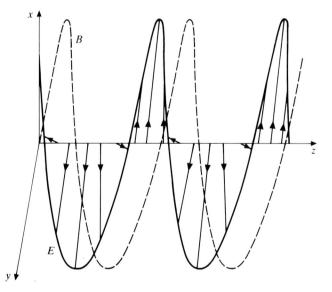

the Doppler effect for electromagnetic radiation is symmetrical in this manner. The result has to be derived using Einstein's Special Theory of Relativity, and so will merely be stated here.

If the source S approaches the observer O with velocity V directed at an angle θ to the line of sight (as shown in fig. 2.6), the Doppler shift is given by

$$v'/v = (1 - V^2/c^2)^{\frac{1}{2}}/(1 - V\cos\theta/c) \qquad (2.26)$$

If $|V/c| \ll 1$, which will always be true in cases of interest to us, this can be approximated as

$$v'/v = 1 + V\cos\theta/c \qquad (2.27)$$

Thus, for example, if a satellite travelling away from an observer on earth with a velocity of $7\,\mathrm{km\,s^{-1}}$ at an angle of $5°$ to the line of sight emits a radio signal at $5\,\mathrm{GHz}$, the signal will be received at a frequency $116\,\mathrm{kHz}$ lower.

2.3 Spectra

Up to this point we have said nothing about the frequency (or wavelength) of the radiation, other than that electromagnetic radiation may, in principle, have any frequency we wish. It will often happen, however, that we wish to describe a particular radiation field in which a number (possibly a continuous distribution) of frequencies is present. This can be done by specifying either the complete waveform, which obviously contains all the necessary information, or the *spectrum* of the radiation – the amplitudes of the various frequency components which are present in the waveform. These two methods are equivalent as we shall see, and it is important to know how to convert from one description to another. The conversion is achieved using the *Fourier transform*, and since this is of great importance in the whole subject of remote sensing it is worth deriving the theory.

Fig. 2.6. The Doppler effect. The source of electromagnetic radiation is located at S, travelling with velocity V. The observer is located at O.

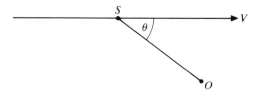

Plate 1. A satellite aerial photograph. The image was recorded by the Metric Camera, carried by the Space Shuttle at an altitude of approximately 250 km, and covers an area of about 190 × 130 km. It shows the Himalaya mountains, including Everest (at the top left, slightly clouded), and the Ganges plain (bottom right). (© *European Space Agency. Reproduced with permission*).

19

Plate 2. False-colour infrared aerial photograph. The image was recorded
from an altitude of approximately 760 m, and covers an area of about 1.1
km square. It shows the Tay reed beds on the Firth of Tay, Scotland.
(Cambridge University Collection: copyright reserved).

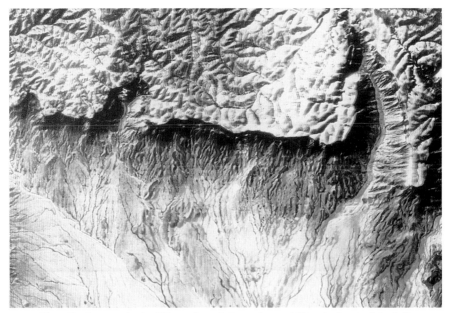

Plate 3. An anaglyph. When viewed through a red filter in front of the left eye and a green filter in front of the right eye, an impression of three-dimensional relief can be obtained. The image was recorded by the SPOT-1 satellite from a height of approximately 830 km, and has a coverage of about 10 × 8 km. It shows part of Djebel Nefoussa, Libya. (© *CNES 1986.*
Reproduced by courtesy of Nigel Press Associates Ltd)

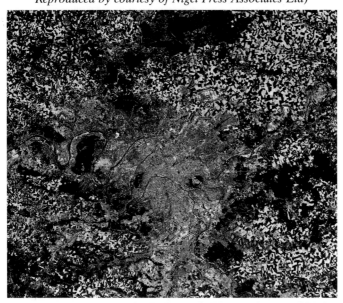

Plate 4. True-colour LANDSAT Thematic Mapper image of Paris. The image was obtained from a height of approximately 700 km, and has a coverage of about 90 × 80 km. (© *European Space Agency. Reproduced with permission)*

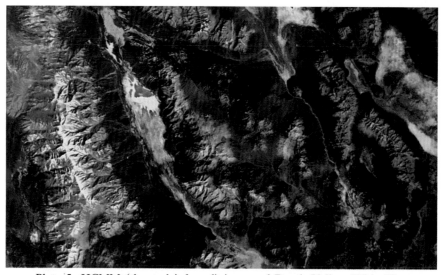

Plate 5. HCMM (thermal infrared) image of Death Valley, USA. The image was acquired on 31 May 1978, during the daytime, and has been colour coded with blue representing the lowest emission of radiation and red the highest. The image has been superimposed on a LANDSAT image of the same area to provide a geographical base. It was obtained from a height of approximately 620 km and has a coverage of about 70 × 100 km. It shows the Panamint (left) and Amargosa (centre) ranges, with Death Valley itself between them. *(Reproduced by courtesy of National Remote Sensing Centre, UK, and the National Aeronautics and Space Administration, USA)*

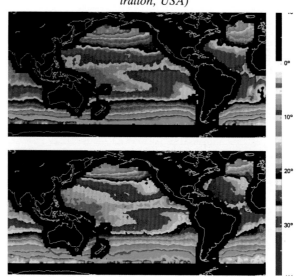

Plate 6. Sea-surface temperature deduced from passive microwave (NIMBUS SMMR) data. The upper figure shows the average temperature for November 1978 to February 1979, the lower figure for July to October 1979. *(Reproduced by courtesy of the National Aeronautics and Space Administration, USA)*

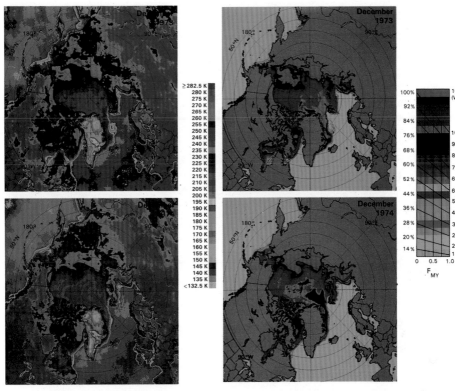

Plate 7. Passive microwave observations (NIMBUS electrically-scanned microwave radiometer) used for the delineation and identification of Arctic sea ice. The figures on the left show the average brightness temperature (at 19.4 GHz) of the sea surface in December 1973 and 1974, and the figures on the right show the corresponding interpretation in terms of sea-ice concentration. The colour scale nomogram indicates the ice concentration (in %) as a function of the fraction F_{MY} of ice which is 'multi-year' (i.e. has survived at least one melt season). *(Reproduced by courtesy of the National Aeronautics and Space Administration, USA, from Parkinson* et al. *1987)*

Plate 8. Bispectral (*L* and *X* band) SAR image of Breisach, West Germany. The image has been colour-coded with *L*-band data represented as blue and *X*-band data as red. The coverage is about 8 by 11 km. The image was obtained from an altitude of approximately 6400 m in July 1984, and clearly shows the river Rhine and the Grand Canal d'Alsace, the town of Breisach, wooded areas, lakes, fields and field boundaries, and the terraced Kaiserstuhl (top right). Note also the very strong signal from the bridge (left centre) across the Rhine between Germany and France. *(Reproduced from Sieber & Noack (1986) by courtesy of the European Space Agency)*

Plate 9. Density slicing. This image shows the thermal infrared channel of part of a LANDSAT image of the Isle of Wight. In the upper part of the figure, the raw data are shown on a 'grey scale'. In the lower part of the figure, different ranges of image DN have been assigned different colours.
(Courtesy of National Remote Sensing Centre, UK)

Plate 10. Multispectral classification. LANDSAT data from the Isle of Wight and Hampshire, England, have been classified as follows: urban areas (blue), woodland and grassland (green), bare fields (brown), crops (yellow), water and unclassified pixels (black). *(Courtesy of National Remote Sensing Centre, UK)*

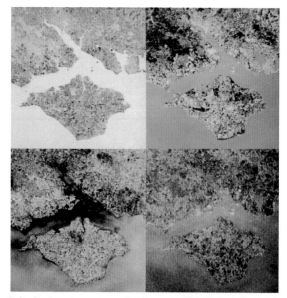

Plate 11. Principal components of an image. This figure shows the first four principal components of a 6-band image of the Isle of Wight and Hampshire, England. The first principal component (top left) contains 88.3% of the total variance of the image. The second component (top right) contains 8.8% of the variance, and the third and fourth components contain 2.3% and 0.3% respectively. *(Courtesy of National Remote Sensing Centre, UK)*

Plate 12. False-colour composite image made by combining the second, third and fourth principal components shown in plate 11. Note the variation in sea colour, caused by suspended sediments, and the way in which distinct classes (see plate 10) have been identified, but note also the extent to which imperfections in the original image have been exaggerated. *(Courtesy of National Remote Sensing Centre, UK)*

Let us suppose that a wave (or any other time-varying quantity) is written as a function of time $f(t)$, and that it is also possible to express it as the sum of components of various angular frequencies ω. If the distribution of frequencies is continuous, the amount of each frequency present can be specified by a density function $a(\omega)$, such that the total amplitude of the components having frequencies in the range $\omega + d\omega$ ($d\omega$ being very small) is $a(\omega)d\omega$. The complex exponential notation simplifies the analysis, so we employ it to write the time-dependence of these components as

$$a(\omega)d\omega \exp(i\omega t)$$

The sum of the contributions of all frequencies will be the integral of this expression, thus

$$f(t) = \int_{-\infty}^{\infty} a(\omega) \exp(i\omega t)\, d\omega \tag{2.28}$$

So far this is merely an assertion. We have neither proved that the distribution $a(\omega)$ uniquely represents $f(t)$, nor shown how to find $a(\omega)$ given $f(t)$. It is beyond our scope to find a rigorous answer to the former problem, so we shall content ourselves with answering the latter problem and relying on physical insight to satisfy ourselves of the former.

If we multiply (2.28) by $\exp(i\omega' t)$, where ω' is an arbitrary angular frequency, we obtain

$$f(t) \exp(i\omega' t) = \int_{-\infty}^{\infty} a(\omega) \exp(i[\omega + \omega']t)\, d\omega$$

Now we integrate this with respect to t, giving

$$\int_{-\infty}^{\infty} f(t) \exp(i\omega' t)\, dt = \int_{-\infty}^{\infty} \int_{-\infty}^{\infty} a(\omega) \exp(i[\omega + \omega']t)\, d\omega\, dt$$

$$= \int_{-\infty}^{\infty} a(\omega) \int_{-\infty}^{\infty} \exp(i[\omega + \omega']t)\, d\omega\, dt$$

Now

$$\int_{-\infty}^{\infty} \exp(i\alpha t)\, dt$$

does not have an analytic form, but it can be shown to be a function of α which is zero everywhere but at $\alpha = 0$, at which it is infinite. The area underneath a graph of the function is however finite, and has a value of 2π. This is written

$$\int_{-\infty}^{\infty} \exp(i\alpha t)\, dt = 2\pi\delta(\alpha)$$

where $\delta(\alpha)$ is the *Dirac delta-function*. Thus we have

$$\int_{-\infty}^{\infty} f(t)\exp(i\omega' t)\, dt = 2\pi a(-\omega')$$

which can be rewritten, by changing the symbols and rearranging the expression, as

$$a(\omega) = \frac{1}{2\pi} \int_{-\infty}^{\infty} f(t)\exp(-i\omega t)\, dt \qquad (2.29)$$

This is very similar to (2.28) and shows that, apart from a change of sign and scale, $a(\omega)$ is obtained from $f(t)$ in exactly the same way as $f(t)$ is obtained from $a(\omega)$. The integral transforms defined by (2.28) and (2.29) are called *Fourier transforms*. It is worth noting that some authors increase the symmetry between (2.28) and (2.29) still further by writing

$$f(t) = \frac{1}{(2\pi)^{\frac{1}{2}}} \int_{-\infty}^{\infty} a(\omega)\exp(i\omega t)\, d\omega$$

$$a(\omega) = \frac{1}{(2\pi)^{\frac{1}{2}}} \int_{-\infty}^{\infty} f(t)\exp(-i\omega t)\, dt$$

Let us apply the Fourier transform to a practical example. Suppose we have a waveform $f(t)$ which consists of a single frequency ω_0 which is turned on for a finite time T (fig. 2.7). What is its spectrum $a(\omega)$? We might think that since we used only one frequency to construct $f(t)$, the spectrum would consist of a single spike or delta-function centred at that frequency. However, this cannot be correct since the spectrum $a(\omega)$ has to

Fig. 2.7. A truncated cosine wave. The Fourier transform of this function is shown in fig. 2.9.

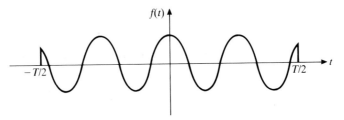

contain *all* the information contained by $f(t)$, including the fact that the waveform drops abruptly to zero for $|t| > T/2$. Using (2.29), then, we find

$$a(\omega) = \frac{1}{2\pi} \int_{-T/2}^{T/2} \cos(\omega_0 t) \exp(-i\omega t)\, dt$$

$$= \frac{1}{4\pi} \int_{-T/2}^{T/2} [\exp(i\{\omega_0 - \omega\}t) + \exp(i\{-\omega_0 - \omega\}t)]\, dt$$

$$= \frac{1}{2\pi} \left[\frac{\sin(\omega_0 - \omega)T/2}{\omega_0 - \omega} + \frac{\sin(\omega_0 + \omega)T/2}{\omega_0 + \omega} \right]$$

This is evidently the sum of two identical functions, each of the form $(\sin x)/x$, centred at frequencies ω_0 and $-\omega_0$. The function $(\sin x)/x$, often called sinc (x), is shown in fig. 2.8. (Note that some authors define sinc (x) as $(\sin \pi x)/(\pi x)$).

Thus the complete spectrum of the waveform whose time dependence was shown in fig. 2.7 is shown by fig. 2.9. It can be seen that the delta-functions which we might have expected at $\omega = \pm \omega_0$ have been broadened by $2\delta\omega$ where

$$\delta\omega \approx 2\pi/T \qquad\qquad (2.30)$$

or $\delta\nu \approx 1/T$. This is in fact a general result of fundamental importance: in order to represent a waveform of length δt, we need a range of frequencies of at least $\pm 1/\delta t$. It is a form of 'uncertainty principle'. Defining exactly what is meant by 'length' and 'range' is not always obvious, so the result as stated in (2.30) can only be applied approximately, although a more exact formulation of the result is possible.

Fig. 2.8. The function sinc(x), defined as $(\sin x)/x$.

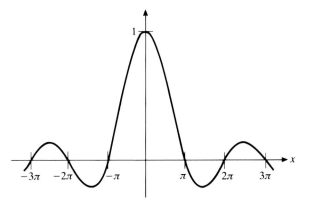

A related result, which is however quite exact, is the *Nyquist sampling theorem*. This states that if a signal is to be sampled discretely, the sampling frequency must exceed a minimum value in order that the signal should be unambiguously reconstructible from the samples. This frequency is the *Nyquist frequency*, and it is twice the bandwidth of the signal. The bandwidth is defined as the range of frequencies v over which the signal spectrum is non-zero. If the signal is undersampled, aliases (to which we shall refer again in chapter 9) are introduced which, amongst other undesirable effects, degrade the signal-to-noise ratio. The practical applications of the Nyquist theorem are many, but it clearly finds an important application in the design of electronic systems in which a signal is first filtered to define a bandwidth, and then sampled at regular intervals. This type of system will be considered in greater detail in chapters 6 and 8.

2.4 Diffraction

In the previous section we discussed the description of a train of waves of limited extent, and developed the theory of Fourier transforms to relate the time and frequency behaviour of any function. In this section we wish to consider the effect of a spatial limitation on the wavefront. This will lead to results which are of fundamental importance in understanding the *resolution* of remote sensing systems and the properties of radar antennas.

We shall begin by considering a one-dimensional restriction of the wavefront. Let us suppose that plane parallel radiation is incident on an aperture as shown in fig. 2.10.

The aperture has an *amplitude transmittance function* $f(y)$, which clearly falls to zero in the opaque parts of the aperture. We wish to

Fig. 2.9. Fourier transform of the function shown in fig. 2.7.

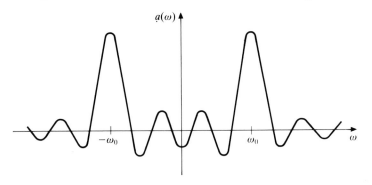

determine the field at the point P. If the distance z is sufficiently large (we shall see how large later), the rays OP and AP may be regarded as being parallel, and AP is shorter than OP by $y \sin \theta$. The phase difference is thus $ky \sin \theta$, where k is the wavenumber of the radiation. If this condition is met, what we are describing is termed Fraunhofer diffraction. The (complex) amplitude at P contributed by an element of the aperture of width dy, located at A, is thus $f(y) \exp(iky \sin \theta) \, dy$. (We are ignoring the reduction of amplitude with distance due to geometrical spreading, as well as one or two other effects.) The total amplitude at P is given by the integral of this expression across the entire aperture, i.e. for all values of y, thus

$$a(\theta) = \int_{-\infty}^{\infty} f(y) \exp(iky \sin \theta) \, dy \qquad (2.31)$$

This is clearly a Fourier transform, although (2.31) is usually called the *Fraunhofer diffraction integral*. In the preceding section we identified time and angular frequency ω as a pair of conjugate variables related through the Fourier transform; here the corresponding variables are distance (y) in the aperture plane, and scaled angle ($k \sin \theta$). Again a form of uncertainty principle holds. If we put $f(y) = 1$ when $|y| < L/2$ and $f(y) = 0$ when $|y| \geq L/2$ to represent a uniform slit of width L, we find on applying the diffraction integral (2.31) that $a(\theta)$ is proportional to $\mathrm{sinc}(kL \sin \theta/2)$. This is a function of the same shape as fig. 2.8, and it first falls to zero when $\sin \theta = \pm 2\pi/kL = \pm \lambda/L$. If L is much greater than the wavelength λ, $\sin \theta$ will be much less than 1 so we can put $\sin \theta \approx \theta$ and

$$\delta\theta \approx \lambda/L \qquad (2.32)$$

Fig. 2.10. Geometry of Fraunhofer diffraction.

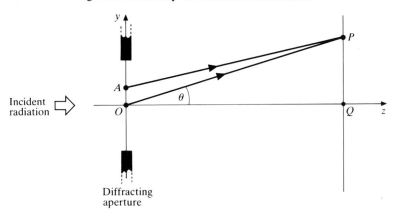

This result corresponds to equation (2.30) and shows that, if a beam of plane parallel radiation passes through an aperture of width L, it will spread into a diverging beam whose width will be of the order of λ/L radians.

If we have a two-dimensional aperture whose amplitude transmittance function is $f(x,y)$, the Fraunhofer diffraction integral of (2.31) becomes

$$a(\theta_x,\theta_y) = \int\limits_{-\infty}^{\infty} \int\limits_{-\infty}^{\infty} f(x,y) \exp(ikx \sin \theta_x) \exp(iky \sin \theta_y) \, dx \, dy$$

$$(2.33)$$

This double integral is in general rather difficult to solve, although there are two special cases which should be mentioned. The first is when $f(x,y)$ can be factorised into two independent parts; $f(x,y) = g(x)h(y)$. This is appropriate to rectangular apertures, and the double integral in (2.33) simplifies to the product of two single integrals of the form of (2.31). The second special case is that of circular symmetry. In this case it is simpler to use polar coordinates. We shall need only one result for general reference, and that is the diffraction pattern of a uniform circular aperture of diameter D. The amplitude of the diffracted field can be shown to be proportional to

$$\frac{J_1(kD \sin \theta_r/2)}{(kD \sin \theta_r/2)}$$

where J_1 is the first-order Bessel function, and θ_r is the radial angle. This function is sketched in fig. 2.11. $J_1(x)$ first falls to zero when $x = 3.832$, so the first zero occurs when $\sin(\theta_r) = 7.66/kD = 1.22\lambda/D$.

Fig. 2.11. The Fraunhofer diffraction pattern of a circular aperture of diameter D. θ_r is the radial angle, i.e. the angle from the normal to the plane of the aperture.

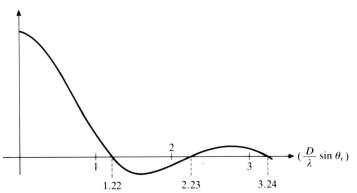

Finally let us return to the remark we made concerning fig. 2.10, that the distance z must be large enough for the two rays OP and AP to be regarded as parallel. How large is this? We assume conventionally that the Fraunhofer description is valid if the phase differences computed using it are accurate to within $\pi/2$ radians. Inspection of fig. 2.10 shows that this is equivalent to putting

$$AQ - OQ < \lambda/4$$

and since OA takes a maximum value of $L/2$ we may use Pythagoras' theorem to derive

$$\{(L/2)^2 + z^2\}^{\frac{1}{2}} - z < \lambda/4$$

Now if $L/2 \ll z$ (which will nearly always be the case) we can use the binomial approximation to reduce this to

$$L^2/8z < \lambda/4$$

or

$$z > L^2/2\lambda = z_F \tag{2.34}$$

The distance z_F is often called the Fresnel distance (after A. Fresnel who made many important discoveries in physical optics in the early nineteenth century), and if the condition (2.34) is not satisfied a more rigorous form of diffraction theory, known as *Fresnel diffraction*, must be used. $z < z_F$ is often called the *near field*, and $z > z_F$ the *far field*.

2.5 Plane boundaries

Next let us review the reflexion and transmission of radiation at a plane boundary between two media

Fig. 2.12 shows a ray in medium 1 making an angle θ_1 with the normal to the boundary with medium 2. In general some of the radiation will be reflected back into medium 1, again at an angle θ_1, and some will be refracted across the interface, to make an angle θ_2 in medium 2. *Snell's law* gives the angle θ_2, as a function of the refractive indices n_1' and n_2' of the two media, through

$$n_1' \sin \theta_1 = n_2' \sin \theta_2 \tag{2.35}$$

We would also like to know the reflexion and transmission coefficients r and t. They are calculated by solving Maxwell's equations at the interface, and the results, in terms of the impedances of the two media, are as follows:

$$
\left.
\begin{aligned}
r_\perp &= \frac{Z_2 \cos \theta_1 - Z_1 \cos \theta_2}{Z_2 \cos \theta_1 + Z_1 \cos \theta_2} \\[1em]
t_\perp &= \frac{2Z_2 \cos \theta_1}{Z_2 \cos \theta_1 + Z_1 \cos \theta_2} \\[1em]
r_\parallel &= \frac{Z_2 \cos \theta_2 - Z_1 \cos \theta_1}{Z_2 \cos \theta_2 + Z_1 \cos \theta_1} \\[1em]
t_\parallel &= \frac{2Z_2 \cos \theta_1}{Z_2 \cos \theta_2 + Z_1 \cos \theta_1}
\end{aligned}
\right\} \qquad (2.36)
$$

These are the amplitude coefficients (usually called the *Fresnel coefficients*), i.e. the electric field is multiplied by a factor of r and t on reflexion and transmission respectively, and the subscripts $\perp$ and $\parallel$ describe whether the E vector of the radiation is perpendicular or parallel to the plane containing the incident and refracted rays. The terms horizontal and vertical polarisation are sometimes used instead of perpendicular and parallel respectively. To understand this notation it is necessary to think

Fig. 2.12. Reflexion and refraction at a plane boundary between two media.

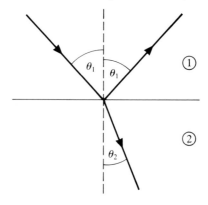

Fig. 2.13. Parallel and perpendicular (vertical and horizontal) polarisations of radiation incident at and reflected from a plane boundary between two media.

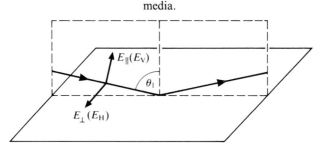

of the surface as being horizontal, and to realise that the 'vertically' polarised radiation merely has a vertical component. If the incidence angle is zero, the distinction between the two kinds of polarisation vanishes. (see fig. 2.13).

The full expressions of (2.36) in the case when both media 1 and 2 are lossy are rather complicated, but if we assume that medium 1 is a vacuum (to which air is normally a sufficiently good approximation) we obtain the following expressions for the *power* reflexion coefficients:

$$
\left.
\begin{aligned}
r_\perp{}^2 &= \frac{(p-\cos\theta)^2 + q^2}{(p+\cos\theta)^2 + q^2} \\[2mm]
r_\parallel{}^2 &= \frac{(\varepsilon'\cos\theta - p)^2 + (\varepsilon''\cos\theta - q)^2}{(\varepsilon'\cos\theta + p)^2 + (\varepsilon''\cos\theta + q)^2}
\end{aligned}
\right\}
\tag{2.37}
$$

where

$$
\begin{aligned}
p &= \{[(\varepsilon' - \sin^2\theta)^2 + \varepsilon''^2]^{\frac{1}{2}} + (\varepsilon' - \sin^2\theta)\}^{\frac{1}{2}}/\sqrt{2}, \\
q &= \{[(\varepsilon' - \sin^2\theta)^2 + \varepsilon''^2]^{\frac{1}{2}} - (\varepsilon' - \sin^2\theta)\}^{\frac{1}{2}}/\sqrt{2}
\end{aligned}
$$

and $\theta = \theta_1$. ε' and ε'' clearly refer to medium 2.

If medium 2 is lossless, the *amplitude* reflexion coefficients become

$$
\left.
\begin{aligned}
r_\perp &= \frac{(n^2 - \sin^2\theta)^{\frac{1}{2}} - \cos\theta}{(n^2 - \sin^2\theta)^{\frac{1}{2}} + \cos\theta} \\[2mm]
r_\parallel &= \frac{n^2\cos\theta - (n^2 - \sin^2\theta)^{\frac{1}{2}}}{n^2\cos\theta + (n^2 - \sin^2\theta)^{\frac{1}{2}}}
\end{aligned}
\right\}
\tag{2.38}
$$

We can see from this that $r_\parallel$ is zero when $\theta = \theta_B$ and $\tan\theta_B = n$. θ_B is called the *Brewster angle*. Parallel (vertically) polarised radiation incident on a surface at the Brewster angle cannot be reflected and so must all be transmitted into the medium. Looked at from another point of view, randomly polarised radiation incident from all directions on a boundary will in general, on reflexion, be partially polarised; and at the Brewster angle will be completely plane polarised. This is the justification for the remark that we made earlier about the degree of polarising being changed by reflexion at a surface.

2.6 Thermal radiation

Thermal radiation is emitted by all objects at temperatures above absolute zero and is, at first or second hand, the radiation which is detected by the vast majority of passive remote sensing systems. In general, a hot object (by which, for the present, we mean one which is not at absolute zero) will distribute its emission over a range of wavelengths and directions, so we shall need to introduce the terms necessary for describing this kind of radiation. This will also be useful in chapter 3, where we consider the reflexion of radiation from rough surfaces and will need much of the same terminology.

Let us consider a small area dA on which radiation is incident from a variety of angles. In particular, let us consider the power which is arriving at dA in the general direction θ to the normal to dA, but in a range of directions which subtend a solid angle $d\Omega$ steradians about that direction (fig. 2.14). We define the *radiance* of the incoming radiation to be L, where

$$d\Phi = L\,dA\,d\Omega \cos\theta \qquad (2.39)$$

is the power incident from that particular direction. From this definition, it follows that radiance is measured in units of $Wm^{-2}sr^{-1}$. If the medium through which the radiation propagates does not absorb or scatter and has a constant refractive index, then L is constant along a ray. The total power falling on dA from all directions is given by integrating (2.39) over 2π steradians (since we are considering radiation arriving from one side only). Thus

$$\Phi = dA \int_{2\pi} L \cos\theta\,d\Omega \qquad (2.40)$$

$$= E\,dA$$

where E is defined by (2.40) and is called the *irradiance* of the radiation field at the surface. It is measured in Wm^{-2}. If we now reverse the directions of the arrows in fig. 2.14, so that dA is emitting radiation instead of receiving it, the term equivalent to E which measures all the radiation leaving dA is called the (radiant) *exitance* M, and the total

Fig. 2.14. Geometrical construction to explain the concept of radiance.

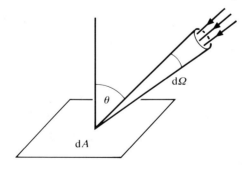

Table 2.1. *Some terms used in radiometry*

Quantity	Symbol	Definition	Unit
Radiant flux	ϕ		W
Radiance	L	$d^2\phi/(\cos\theta\,d\Omega\,dA)$	$\mathrm{W\,m^{-2}\,sr^{-1}}$
Irradiance (in)	E	$d\phi/dA$	$\mathrm{W\,m^{-2}}$
Radiant exitance (out)	M	$d\phi/dA$	$\mathrm{W\,m^{-2}}$
Radiant intensity (out)	I	$d\phi/d\Omega$	$\mathrm{W\,sr^{-1}}$

power emitted by the whole source in a given range of directions is called the *radiant intensity I*, defined by

$$\Phi_{\mathrm{total}} = I\,d\Omega \qquad (2.41)$$

where Φ_{total} is the total power radiated by the source into a cone of solid angle $d\Omega$, in the specified direction. These quantities are summarised in table 2.1.

In most cases, and certainly in the case of thermal radiation, the radiation will be distributed in wavelength (or frequency) as well as in direction. In this case we redefine the quantities we have just introduced *spectrally*, that is, over an interval of wavelength λ or frequency v as appropriate. The usual way of indicating that a quantity is defined spectrally is to give it an appropriate subscript of λ or v. Thus (2.39) may be rewritten as

$$d\Phi = L_\lambda \cos\theta\,dA\,d\Omega\,d\lambda$$
$$= L_v \cos\theta\,dA\,d\Omega\,dv$$

where L_λ and L_v are the spectral radiances. It is clear from this that L_λ and L_v do not have the same units, being measured in $\mathrm{W\,m^{-3}\,sr^{-1}}$ and $\mathrm{W\,m^{-2}\,sr^{-1}\,Hz^{-1}}$ respectively. It also follows that L_λ and L_v are related by

$$L_\lambda/L_v = |dv/d\lambda| = c/\lambda^2 = v^2/c \qquad (2.42)$$

where c is the velocity of light.

We are now in a position to describe thermal radiation. As we remarked before, all objects above absolute zero emit radiation in a continuous spectrum, i.e. over a range of wavelengths. If we make a closed cavity with opaque walls at a temperature T (absolute), the radiation inside it is known as *black-body* radiation. The spectral radiance of this radiation was calculated by Planck during the early years of the twentieth century, using quantum mechanics (see e.g. Longair 1984). It is

Table 2.2. *Integral of the Planck distribution function*

The spectral radiance of a black body may be written in a normalised, dimensionless form as

$$g(x) = \frac{15}{\pi^4} \frac{x^3}{e^x - 1}$$

where $x = hc/(\lambda kT)$. The table gives values of

$$f(z) = \int_0^z g(x)\,dx$$

z	$f(z)$	z	$f(z)$
0.10	0.00005	1.6	0.11023
0.12	0.00009	1.8	0.14402
0.14	0.00013	2.0	0.18115
0.16	0.00020	2.5	0.28403
0.18	0.00028	3.0	0.39302
0.20	0.00038	3.5	0.49938
0.25	0.00073	4.0	0.59703
0.30	0.00124	4.5	0.68251
0.35	0.00193	5.0	0.75453
0.40	0.00282	6.0	0.86016
0.45	0.00394	7.0	0.92443
0.50	0.00529	8.0	0.96084
0.60	0.00879	9.0	0.98039
0.70	0.01341	10.0	0.99045
0.80	0.01923	12.0	0.99788
0.90	0.02629	14.0	0.99956
1.0	0.03462	16.0	0.99991
1.2	0.05506	18.0	0.99998
1.4	0.08040	20.0	1.00000

Example: To calculate the exitance of a black body at 300 K between 10 and 14 μm. At 14 μm, $z = 3.43$ so $f(z) \approx 0.4845$, and at 10 μm, $z = 4.80$ so $f(z) \approx 0.7257$. Thus the fraction of the total exitance emitted between these wavelengths is approximately 0.241, and since the total exitance is $5.67 \times 10^{-8} \times 300^4 \, W\,m^{-2}$, the required answer is $1.1 \times 10^2 \, W\,m^{-2}$.

$$L_v = \frac{2h\nu^3}{c^2} (e^{h\nu/kT} - 1)^{-1} \qquad\qquad (2.43)$$

which may also be expressed as

$$L_\lambda = \frac{2hc^2}{\lambda^5} (e^{hc/\lambda kT} - 1)^{-1} \qquad\qquad (2.44)$$

using (2.42). In these expressions, h is Planck's constant and k is the Boltzmann constant. The general form of this function is shown in fig. 2.15. Note in particular the steep rise at short wavelengths, and the long tail at long wavelengths. Since this function cannot easily be integrated, it is tabulated in table 2.2 to facilitate numerical calculations.

The radiation inside a closed cavity may not seem to be particularly interesting or relevant, but we may observe it by making a small hole in the cavity and letting some of it escape. In this case, (2.43) or (2.44) describes the radiation emerging from the hole, and from any black body (i.e. perfect emitter of thermal radiation) at the temperature T.

For isotropic radiation, L_λ is independent of direction so the spectral radiant exitance is given by

$$M_\lambda = L_\lambda \int \cos \theta \, \mathrm{d}\Omega$$
$$= \pi L_\lambda$$

The total outgoing radiance of a black body is given by integrating (2.44) with respect to wavelength to give

$$L = \int_0^\infty L_\lambda \, \mathrm{d}\lambda = \frac{2\pi^4 k^4}{15c^2 h^3} T^4$$

and the total radiant exitance is thus

$$M = \int_0^\infty M_\lambda \, \mathrm{d}\lambda = \frac{2\pi^5 k^4}{15c^2 h^3} T^4 = \sigma T^4 \tag{2.45}$$

where $\sigma = 2\pi^5 k^4/15c^2 h^3 = 5.67 \times 10^{-8} \, \mathrm{Wm}^{-2} \mathrm{K}^{-4}$ is called the Stefan–Boltzmann constant, and (2.45) is called Stefan's law. It shows how much

Fig. 2.15. Black-body radiances according to the Planck law. The graphs show $\log_{10} L_\lambda$ in units of $\mathrm{W\,m}^{-3}\,\mathrm{sr}^{-1}$ plotted against $\log_{10}\lambda$ in metres.

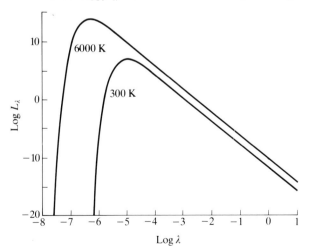

power is emitted by a black body at temperature T, over all wavelengths. If we want to know how this power is distributed in wavelength, we can of course use the Planck radiation formula (2.44), but it may be sufficient merely to know the wavelength λ_{max} at which L_λ reaches a maximum. This is found by differentiating (2.44), which shows that

$$\lambda_{max} = c_W / T \qquad\qquad (2.46)$$

where c_W is a constant whose value is about 2.898×10^{-3} Km. Equation (2.46) is known as Wien's law. For example, the sun is a reasonably good approximation to a black body with a temperature of 5800 K, so the maximum spectral radiance occurs at a wavelength of 0.5 μm, in the visible region as we expect. If on the other hand we consider a black body at a temperature of 280 K, which is fairly typical of temperatures on the earth's surface, $\lambda_{max} = 10.3$ μm which is in the thermal infrared region of the spectrum.

We remarked before that a small hole in the wall of a cavity behaves as a black body. This is not a particularly plausible model for real materials, so we introduce the idea of the *emissivity* ε to relate the actual radiance of a body at a temperature T to the black-body value. (Note that emissivity and dielectric constant have the same symbol epsilon, which is potentially somewhat confusing. This usage is too well established, however, for us to introduce a different notation, and we will rely on context, or an explicit statement, to differentiate between them.) The emissivity is often dependent on wavelength, so it must be defined spectrally. Thus the spectral radiance of a real object is

$$L_\lambda = \varepsilon_\lambda L_{\lambda,p}$$

where we have now written $L_{\lambda,p}$ for the black-body radiance (the subscript 'p' stands for Planck). A simple thermodynamic argument shows that a body which is a good emitter (high ε) must also be a good absorber of radiation – in fact the two factors must be equal. We can see this quite easily by realising that any body at temperature T must be in equilibrium with black-body radiation whose spectrum corresponds to the same temperature. If the body absorbs better than it emits, say, then it will heat up, and thus cannot in fact be at equilibrium. Thus the reflectivity is given by $1 - \varepsilon$. It also follows from this argument that the emissivity ε must lie between zero and one.

We can use the results we have developed to characterise, approximately, sunlight. To a fairly good approximation the sun can be taken to be a *grey body* (i.e. it has a constant ε_λ over the range of emission) with a temperature T of about 5800 K and an emissivity ε of 0.99. It has a radius r of 6.96×10^8 m and it is located at a distance D from the earth of 1.496×10^{11} m (see appendix). Putting these data together with what we have learnt about black-body radiation, we may calculate:

Total emitted power $= 4\pi r^2 T^4 \sigma \varepsilon$	$= 3.9 \times 10^{26}$ W
Irradiance at earth $= r^2 T^4 \sigma \varepsilon / D^2$	$= 1.37 \times 10^3$ Wm^{-2}
Radiance of the sun's disc seen from earth $= T^4 \sigma \varepsilon / \pi$	$= 2 \times 10^7$ Wm^{-2}sr^{-1}
Wavelength above which 1% of the total power is emitted $= 3.9\,\mu$m	
Wavelength below which 1% of the total power is emitted $= 0.25\,\mu$m	

We see that the limits embracing 98% of the sun's power are 0.25 to $3.9\,\mu$m, covering the near UV, visible and near IR bands. This then is the region of the spectrum in which we expect to be able to detect scattered sunlight.

The mean sea and sea-level air temperatures at 50° (N or S) latitude are about 7 °C or 280 K. The temperatures of nearly all other naturally occurring objects on or near the surface of the earth do not differ greatly from this. A black body at 280 K radiates with $\lambda_{max} = 10\,\mu$m, in the thermal infrared. Thus unless such an object has an exceptionally low emissivity we expect it to radiate in the thermal IR band. We know that the Planck distribution cuts off rather sharply at the low-wavelength end but has a long tail at the high-wavelength end. Thus we expect to find virtually no radiation emitted by such bodies at wavelengths shorter than a few microns, but significant amounts may be emitted at even quite long wavelengths. As we shall see in chapter 3, the atmosphere effectively absorbs wavelengths from about 20 μm to 2 mm, so the most likely part of the spectrum in which to look for radiation is the millimetre to centimetre (i.e. microwave) region. In fact, we find that there is no insuperable difficulty in building receivers sensitive enough to pick up thermal radiation in this band, and this kind of technique is called passive microwave radiometry (PMR). It is discussed in chapter 6.

Problems

1. The flux density of sunlight near the earth is about 1370 Wm^{-2}. Find the rms values of the electric and magnetic fields.
2. Sea water has a conductivity of ≈ 4 Sm^{-1}, and the real part of its permittivity below about 100 MHz may be taken as 88.2. Find the absorption length for electromagnetic waves at 100 MHz, and at 100 kHz.
3. Calculate by how much slower a pulse of light with free-space

wavelength $0.5\,\mu m$ will travel than a pulse of near-infrared radiation of free-space wavelength $1.0\,\mu m$, if both pulses travel through air at STP with a water vapour partial pressure of $1500\,Pa$ (see appendix).

4. Estimate the Fresnel distance for (a) an optical system with an objective diameter of $0.2\,m$, and (b) a microwave system with an 'objective' (antenna) diameter of $10\,m$.

5. Estimate the real and imaginary parts of the refractive index of pure water (at $20\,°C$) at (a) $100\,MHz$, and (b) $10\,GHz$. Calculate the power reflexion coefficients at each of these frequencies for horizontally and vertically polarised radiation incident at an angle of $50°$ to the vertical.

6. What is the ratio of the spectral radiances of black bodies at $300\,K$ and $6000\,K$ at (a) $1\,GHz$, (b) $1000\,GHz$, (c) $1\,\mu m$, and (d) $0.1\,\mu m$?

7. Show that, for a black body, the wavelength at which L_v is maximum is about 1.76 times greater than the wavelength at which L_λ is maximum at the same temperature.

8. What fraction of the sun's radiation is emitted between 8 and $14\,\mu m$? What would be the apparent temperature of the sun measured by an instrument which integrates the radiation over this range of wavelengths and assumes the result to be proportional to $T^{4\cdot6}$? (Assume the instrument to be calibrated for objects near $300\,K$.)

3

Interaction of radiation with the surface and atmosphere

3.1 Introduction

In chapter 2 we reviewed the generation of electromagnetic radiation by the black-body process, and the propagation of radiation through homogeneous media. However, our consideration of how electromagnetic radiation interacts with a surface was restricted to the case where there is a plane boundary separating two media. In this chapter, we will discuss the way in which radiation interacts with real surfaces, which will lead to a discussion of how such surfaces are characterised, and how their reflexion properties are quantified using the bidirectional reflectance distribution function.

As was pointed out in the preface, the succession of chapters in this book is intended to follow as closely as possible the flow of information in remote sensing. After the considerations of the previous chapter have been made, we will have, so to speak, an electromagnetic wave modulated in some way by the surface being sensed. The information carried by this wave will be located near the surface, and has yet to travel through some or all of the earth's atmosphere in order to reach the detector. It is clear that the atmosphere may introduce further modulation of the wave, and we must therefore consider the interaction of electromagnetic waves with the atmosphere. This is the substance of the later parts of the present chapter.

3.2 Reflexion from rough surfaces

All active remote sensing systems, as well as those passive systems which detect reflected sunlight, involve the reflexion of radiation from the surface of interest. However, as we saw in chapter 2, the thermal emissivity of a surface is directly related to its reflectance properties, so even in the case of passive microwave and thermal infrared remote sensing the roughness-related reflexion properties are of fundamental importance.

In this section we will define the reflectance and roughness properties of a surface, and then show how they are related. In particular, we will distinguish two special cases, the *specular reflector*, which is perfectly smooth, and the *Lambertian reflector*, which is 'perfectly rough'. Of course both of these are idealised and never realised in practice, although they may well suffice in particular circumstances. It should be noted that, for simplicity, the treatment of rough-surface scattering in this section will avoid any discussion of polarisation effects. These are generally unimportant in a consideration of visible or infrared wavelength systems, although they are significant for microwave systems.

3.2.1 Description of surface scattering

Let us begin by developing some of the terms needed to describe the reflexion properties of a surface. (Note: the treatment given in this section is amplified in a number of works. Swain & Davis (1978) is a particularly useful source of information on quantitative descriptions of surface scattering in remote sensing, and Schanda (1986) also provides a helpful treatment.)

Fig. 3.1 shows a well collimated beam of radiation of flux density $F(\mathrm{Wm}^{-2})$, measured perpendicularly to the direction of propagation,

Fig. 3.1. Radiation, initially of flux density F, is incident at angle θ_0 on an area $\mathrm{d}A$ and is then scattered into solid angle $\mathrm{d}\Omega_1$ in the direction θ_1. The azimuthal angles ϕ_0 and ϕ_1 are omitted for clarity.

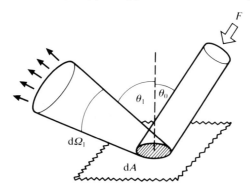

incident on a surface at angle θ_0. The irradiance E at the surface is clearly given by $F\cos\theta_0$. A component of this radiation is scattered into solid angle $d\Omega_1$, in a direction specified by the angle θ_1. θ_0 is often called the *incidence angle*, and its complement, $\pi/2-\theta_0$, the *depression angle*. (At least, this is true over a flat surface. If the surface curves, somewhat less convenient definitions have to be employed.) For simplicity, fig. 3.1 does not show the azimuthal angles ϕ_0 and ϕ_1. Let us suppose that the outgoing radiance of the surface as a result of this illumination is $L_1 \, (\text{Wm}^{-2}\text{sr}^{-1})$ in the direction (θ_1,ϕ_1). The *bidirectional reflectance distribution function* (BRDF) R is defined as

$$R=L_1/E \tag{3.1}$$

Thus R has the units of sr^{-1}, although it is common in radar work to refer to the (unitless) bistatic scattering coefficient γ,

$$\gamma=4\pi R\cos\theta_1 \tag{3.2}$$

Note that the BRDF is also commonly represented by the symbols f and ρ. R is a function of the incident and scattered directions, so in principle we should write it as $R(\theta_0,\phi_0,\theta_1,\phi_1)$. This notation is useful in that it allows us to state the reciprocity relation obeyed by the BRDF, namely that $R(\theta_0,\phi_0,\theta_1,\phi_1)=R(\theta_1,\phi_1,\theta_0,\phi_0)$, but for compactness we shall assume the variables to be implied. In the majority of cases the surface will lack azimuthally-dependent features so that the dependence of R on ϕ_0 and ϕ_1 will simplify to a dependence on $(\phi_0-\phi_1)$, and often the azimuthal dependence may be neglected altogether.

The *reflectivity r* of the surface is a function only of the incidence direction, and it expresses the ratio of the total power scattered to the total power incident. It is thus given by

$$r(\theta_0,\phi_0)=M/E$$

where M is the radiant exitance, whence

$$r(\theta_0,\phi_0)=\int R\cos\theta_1 \, d\Omega_1$$

$$=\int_{\theta=0}^{\pi/2}\int_{\phi=0}^{2\pi} R\cos\theta_1 \sin\theta_1 \, d\theta_1 \, d\phi_1 \tag{3.3}$$

The subscript 1, as before, denotes the scattered direction. The reflectivity is also commonly known as the *albedo* (from the Latin for 'whiteness') of the surface, and it is related to the emissivity through

$$r=1-\varepsilon$$

Table 3.1. *Typical values of albedo*
(Integrated over the visible waveband for normally
incident radiation.) (Mostly after Schanda, 1986.)

Material	Albedo (%)
water (naturally occurring)	1–10
water (pure)	2
forest	5–10
crops (green)	5–15
urban areas	5–20
grass	5–30
soil	5–30
cloud (low)	5–65
lava	15–20
sand	20–40
ice	25–40
granite	30–35
cloud (high)	30–85
limestone	35–40
snow (old)	45–70
snow (fresh)	75–90
global average	≈ 35

We can also define the *diffuse* (or *hemispherical*) *albedo* r_d as the average of r over the hemisphere of possible incidence directions. In this case it represents the ratio of the total scattered power to the total incident power when the latter is distributed isotropically. The contribution dE to the irradiance from the direction (θ_0, ϕ_0) is $L_0 \cos \theta_0 \, d\Omega_0$, where L_0 is the incident radiance, and for isotropic illumination L_0 is constant. It follows that

$$r_d = \frac{1}{\pi} \int_0^{\pi/2} \int_0^{2\pi} r(\theta_0, \phi_0) \cos \theta_0 \sin \theta_0 \, d\theta_0 \, d\phi_0 \qquad (3.4)$$

If the albedo has no azimuthal dependence, this becomes

$$r_d = 2 \int_0^{\pi/2} r(\theta_0) \cos \theta_0 \sin \theta_0 \, d\theta_0$$

In general, if the term 'albedo' is used without qualification, it is usually this albedo which is meant.

The diffuse albedo is a function of neither the incident nor the scattered directions, and represents the extent to which the surface scatters incident radiation. For a surface which scatters all of the incident radiation this will be unity. Table 3.1 gives the diffuse albedo of various materials, averaged over the visible waveband. In general, of course, all of the measures of reflectance which we have introduced here can be defined spectrally, and we shall return to the topic of spectral reflectance later.

There are two important limiting cases of rough surface. The first is the *specular scatterer*, which has the property that if radiation is incident in the direction (θ_0, ϕ_0), it will be scattered only into the direction $\theta_1 = \theta_0$, $\phi_1 = \phi_0 - \pi$. Thus the BRDF must contain a delta-function term, shown schematically in fig. 3.2.

The other important limiting case which, like the specular surface, is an idealisation sometimes useful in practice, is the *Lambertian scatterer*. This has the property that, for any illumination which is uniform across the surface, the outgoing radiance is isotropic. This is also illustrated schematically in fig. 3.2. A perfect Lambertian surface will scatter all the radiation incident upon it, so that the radiant exitance M will be equal to the irradiance E and the diffuse albedo will be unity. Writing

$$M = \int_{2\pi} L_1(\theta_1, \phi_1) \cos \theta_1 \, d\Omega_1$$
$$= \pi L_1 \text{ if } L_1 \text{ is constant,}$$

Fig. 3.2. Schematic illustration of different types of surface scattering. The lobes are *polar diagrams* of the scattered radiation. The length of an imaginary line joining the point S (where the radiation is incident on the surface) to the lobe is proportional to the amount of radiation scattered in the direction of the line. (a) specular; (b) quasi-specular; (c) Lambertian; (d) quasi-Lambertian; (e) complex. Note the increased scattering in the specular direction from most real surfaces.

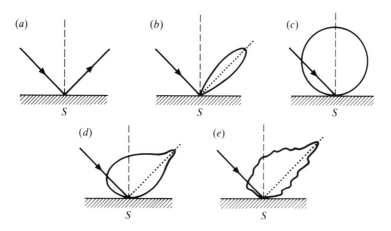

and recalling that $R = L_1/E$, we see that for a perfect Lambertian surface

$$R = 1/\pi \tag{3.5}$$

We have said that the specular and Lambertian scatterers represent idealised, extreme forms of behaviour which are seldom realised in practice. However, it may often happen that the scattering from a real surface bears enough similarity to the ideal case for it to be described as *quasi-specular* or *quasi-Lambertian*.

The behaviour of real scattering surfaces is often specified, not by using the BRDF, but instead by measuring the *bidirectional reflectance factor* (BRF). This is defined as the ratio of the flux scattered into a given direction by a surface under given conditions of illumination to the flux scattered in the same direction by a perfect Lambertian scatterer under identical conditions. The utility of this function is that surfaces can be manufactured which have a BRF very close to unity for a fairly wide range of wavelengths and of incidence and scattering angles. The most common materials are barium sulphate which, as a pressed powder and for θ less than $45°$, has a BRF greater than 0.99 for wavelengths between $0.37\,\mu m$ and $1.15\,\mu m$, and magnesium oxide, which has BRF greater than 0.98 over roughly the same range of conditions.

3.2.2 The Rayleigh criterion

We have distinguished between the behaviour of a perfectly smooth surface, and a Lambertian surface which is in one sense perfectly rough. It is clear that in order to assess which of these forms of behaviour provides the better model of a real surface, some measure of roughness must be adopted. That which is usually adopted is the *Rayleigh criterion*.

Consider the diagram of fig. 3.3, in which radiation is incident on and reflected from a surface irregularity of height Δh, at an angle θ_0. It is clear that the path difference between the scattered ray and a ray which is reflected at the same angle from a height $\Delta h = 0$ is $2\Delta h \cos\theta_0$, and thus the phase difference $\Delta\phi$ is given by

$$\Delta\phi = 4\pi\Delta h \cos\theta_0/\lambda,$$

where λ is the wavelength. A surface can be defined as smooth enough for scattering to be specular if $\Delta\phi$ is less than some arbitrarily defined value of the order of 1 radian. The conventional value is $\pi/2$; this is called the Rayleigh criterion. Thus for a surface to be smooth according to this criterion,

$$\Delta h \cos\theta_0/\lambda < 1/8 \tag{3.6}$$

Note that other criteria, such as $\pi/8$, have also been adopted for the value of $\Delta\phi$ at which a surface becomes effectively smooth. A common definition which provides for the possibility of some intermediate condition between rough and smooth is that if $\Delta\phi$ is greater than $\pi/2$ the surface is rough, and if $\Delta\phi$ is less than $4\pi/25$, it is smooth.

Equation (3.6) evidently dictates that for a surface to be effectively smooth at normal incidence, any irregularities must be less than about $\lambda/8$ in height. Thus for a surface to give specular reflexion at optical wavelengths ($\lambda = 0.5\,\mu\text{m}$, say), Δh must be less than about 60 nanometres. This is a condition of smoothness likely to be met only in certain man-made surfaces such as sheets of glass and metal. On the other hand, if the surface is to be examined using VHF radio waves (say $\lambda = 3\,\text{m}$), Δh need only be less than about 0.4 metres, a condition which could be met by a number of naturally occurring surfaces.

A further aspect of (3.6) is that the restriction on Δh for a surface to reflect effectively specularly becomes less severe as the incidence angle θ_0 is increased. Thus at glancing incidence (large θ_0) a surface may appear quite smooth, whereas at $\theta_0 = 0$ it appears rough. This fact is familiar to anyone who has had to endure the glare of reflected sunlight from a low sun over an ordinary road surface. Although the scattering is by no means specular, the component of the BRDF in the specular direction is greatly enhanced.

3.2.3 *Intermediate cases*

We have derived the BRDF for the simple limiting cases of perfectly smooth and perfectly rough surfaces but, as we remarked before, these are

Fig. 3.3. The Rayleigh criterion. Radiation is specularly reflected at an angle θ_0 from a surface whose r.m.s. height deviation is Δh. The difference in the lengths of the two rays is $2\Delta h\cos\theta_0$.

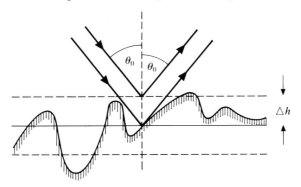

idealisations which are seldom if ever realised in practice. We shall now outline some of the important ways in which the BRDF is calculated in real cases of interest. This section can of course treat this vast and important subject to only a limited extent. The reader who wishes to pursue it in greater depth is recommended to study, for example, the books by Beckmann & Spizzichino (1963), Colwell (1983), and Tsang *et al.* (1985).

3.2.3.1 The small perturbation method

The most constructive way to begin to look at this problem is through the small perturbation method. This is essentially a Fraunhofer diffraction approach to rough scattering, in which the interaction of the incident field with the surface is used to calculate the outgoing field in the vicinity of the surface. This field can be regarded as having been produced from a uniform incident field by a screen which changes both amplitude and phase, and the far-field radiation pattern is obtained by calculating the Fraunhofer diffraction pattern of this 'screen'.

In order to see how this is applied, but without becoming too deeply immersed in mathematical detail, let us consider a one-dimensional surface with a unit diffuse albedo (i.e. all radiation is reflected by it). Let us further assume that the height of this surface above some datum is given by $h(x)$, and that radiation is incident vertically on this surface, and scattered not too far from the vertical (fig. 3.4).

Let us now consider the radiation field, arising as a result of reflexion, on $z=0$ (which for convenience we shall take as being just above the highest point on the surface). It is clear that, relative to a point $h=0$, the radiation reflected from a point at $h(x)$ has travelled an extra distance $-2h$, and has thus suffered a phase delay of $-4\pi h/\lambda$. Thus we can write the radiation (electric) field at $z=0$ as

$$E(0) = \exp(4\pi i h/\lambda),$$

which we can expand as a power series;

$$E(0) = 1 + \frac{4\pi i h}{\lambda} - \left[\frac{4\pi}{\lambda}\right]^2 \frac{h^2}{2} + \cdots$$

The fair field radiation distribution will, as we have seen in chapter 2, be given by the Fraunhofer diffraction integral of $E(0)$. If we write the diffraction integral of $h(x)$ as $H(\theta)$, then

$$E(\text{far field}) = 2\pi\delta(\theta) + \frac{4\pi i}{\lambda} H(\theta) - \frac{1}{2}\left[\frac{4\pi}{\lambda}\right]^2 H(\theta)*H(\theta) + \cdots$$

where the symbol * denotes convolution. The delta-function evidently represents specular reflexion, and the higher terms the rough scattering.

We can note two important facts from this analysis, without any further mathematical development of it. The first is that if $4\pi\Delta h/\lambda \ll 1$, where Δh is a measure of the height of typical roughness elements (e.g. it could be the square root of the variance of the surface height), all terms except the specular term may reasonably be ignored. This is of course directly related to the Rayleigh criterion which we derived earlier.

If the surface is somewhat too rough for the scattering to be considered specular, but nevertheless we have $4\pi\Delta h/\lambda < 1$ so that any term in the expansion is smaller than the previous term, we may regard the rough scattering as being dominated by the term in $H(\theta)$. If this is not the case then the situation is too complicated to be dealt with by this method, but if it is true then the scattering is known as *Bragg scattering*. A somewhat more general description of Bragg scattering is that the scattered wave amplitude is proportional to the Fourier component of the surface height distribution whose wavevector matches the change in the wavevector of the radiation. For example, in one dimension for simplicity, let us consider radiation incident at the angle θ_0 to the surface normal and scattered at the angle θ_1. The component of the wavevector parallel to the surface changes by $\Delta k_s = 2\pi/\lambda(\sin\theta_1 - \sin\theta_0)$, and the scattered amplitude will be proportional to the component of the surface height distribution with spatial frequency Δk_s. The Bragg scattering mechanism is thought to be largely responsible for the reflexion of microwave radiation from small-scale (of the order of 1 centimetre) roughness on water surfaces,

Fig. 3.4. The small perturbation model of rough-surface scattering (simplified). The surface has the equation $z = h(x)$. Radiation is incident at zero angle, and the scattering angle θ_1 is assumed to be small.

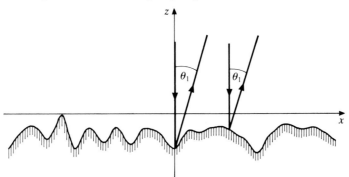

especially where the structure of this roughness contains a dominant spatial frequency, in which case the Bragg scattering is said to be resonant (see Valenzuela, 1978).

3.2.3.2 *The facet model*

A second important model for scattering from rough surfaces is called the *facet model*. Here, the surface is imagined to be composed of planar facets each of which is tangential to the true surface. The criteria which must be satisfied in choosing these facets are that they be much more than one wavelength across (so that diffraction effects do not dominate), and that the true and modelled surfaces do not differ in height anywhere by more than about half a wavelength (so that the phase errors so incurred are not excessive). Provided that these conditions are met, each facet may be considered a specular reflector, and its contribution to the total BRDF may be calculated geometrically as a function of orientation. Although we have said that each facet scatters specularly, the reflected beams will in fact diverge slightly as a consequence of diffraction at the finite facet. The intensity scattered into a given direction is thus, roughly speaking, proportional to the number of facets oriented to within the diffraction angle of the angle required for specular reflexion. This is illustrated schematically in fig. 3.5.

Before we leave the facet model we ought to examine under what circumstances it is valid, as we did for the small perturbation model. It is clear that the surface must be such that we can find facets whose size is much greater than the wavelength λ, and whose deviation from the true surface is much less than λ. This is in fact a restriction on the curvature of the surface. Again, we can model this in one dimension for simplicity. Let us assume that the surface has a radius of curvature R (see fig. 3.6). If the

Fig. 3.5. The facet model of rough surface scattering. The surface is modelled as a series of plane facets which are tangential to the true surface (shaded). Each facet gives quasi-specular (diffraction limited) scattering, represented by polar diagram lobes in the same manner as in fig. 3.2.

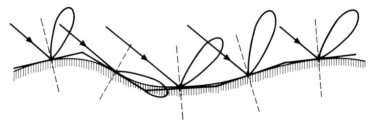

facet size is $2w$, the angle subtended by it at the centre of curvature is 2ψ where ψ is approximately w/R. (We are assuming that $\psi \ll 1$, which we shall shortly see must be true.) The maximum deviation x of the facet from the surface is then $R(\sec\psi - 1)$ which is approximately $R\psi^2/2$, so R is approximately $w^2/2x$. Let us put, for sake of argument, $w \geq \lambda$ and $x \leq \lambda/4$. The condition then becomes $R \geq 2\lambda$. I.e. in general, the radius of curvature must be a few wavelengths or greater. This is evidently a less restrictive condition than for the small perturbation method, although again there will be cases where it is not satisfied.

Another condition on the validity of the facet model is that the incidence or scattering angles should not be large enough that one part of the surface obscures ('shadows') another. If this occurs in practice, it can be taken care of either by modifying the model to allow for it, or by specifying that the model is valid only up to some maximum angle.

We must leave the mathematical modelling of reflexion from rough surfaces at this point. We have described, in a simplified way, the two most important models, but the field is a vast and complicated one in which active research still continues.

3.2.4 Spectral reflectance

Up to this point, in considering scattering from rough surfaces, we have discussed only the way in which reflectance depends on the incidence and viewing directions, and have ignored the effect of wavelength. (This remark requires a slight qualification, since wavelength enters as a factor in the Rayleigh roughness criterion and the small perturbation method.)

Fig. 3.6. Estimation of the validity of the facet model. The required conditions are that $w \gg \lambda$ and $x \ll \lambda$, from which it can be shown that the radius of curvature $R \gg \lambda$.

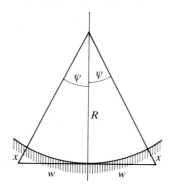

Naturally all of the measures of 'reflecting ability' defined in 3.2.1 may be defined spectrally, in the manner of section 2.6, and indeed this is a particularly important carrier of information about the material being sensed. The spectral reflectance (which we use as a general term to denote reflecting ability under certain but unspecified conditions of incidence and viewing angles) of a target is also a function with which we are familiar, since we frequently apply the principle of identifying objects by their colour (as well as by other properties, of course).

Different target materials have, in general, spectral reflectance curves of different shapes, and this forms a basis for identifying the material from remotely sensed data. The shape of the spectral reflectance curve of a given material is often called its spectral signature. We can thus see the advantage of using a multifrequency remote sensing system, which records data in a number of spectral bands instead of just one. The automatic classification of such data is discussed in chapter 10, but we may note here that the human eye/brain system performs the task instinctively on, for example, normal colour photographs, interpreting green as healthy forest, turquoise as clear water, light green as silty water and so on.

Fig. 3.7 shows the simplified spectral albedo of various materials in the visible and near infrared wavebands of the electromagnetic spectrum. It illustrates the principle we have just described, for it is clear that, for example, a single-channel system operating at about 1.2 μm (in the near infrared) would have considerable difficulty in discriminating between a number of materials, whereas a system with three or four channels between 0.5 and 1.2 μm could extract substantially more information. Before leaving Fig. 3.7, it is useful to remark on some of the principles governing the shapes of these curves.

The albedo of pure *water* is very small (0.02 in the visible region), and is almost constant over this range of wavelengths. It is determined by equations (2.36) from the refractive index of water. The albedo of *vegetable matter*, on the other hand, is governed by the presence of absorbing *pigments* (which produce the small peak in the visible band), and by multiple internal reflexions at interfaces between hydrated cell walls and intercellular air spaces (which produce the high albedo in the near infrared, between about 0.7 and 1.3 μm – see Curran 1985). If the vegetation is diseased, the cell wall structure is damaged and the high albedo between 0.7 and 1.3 μm is reduced, providing a diagnostic remote sensing technique for assessing the health of crops.

The most important of the absorbing pigments is chlorophyll, which has absorption maxima at 0.45 and 0.65 μm, and in consequence there is a local maximum in the spectral reflectance at about 0.55 μm. This is the explanation for the green colour of much vegetable matter, and also explains why unhealthy 'green' vegetation has a yellowish colour. The other important botanical pigments are carotene and xanthophyll (which give orange–yellow reflectance spectra), and the anthocyanins (red–violet). These latter pigments are dominant in the autumn, when the chlorophyll decomposes in many species, and give rise to the spectacular colours of autumn leaves. (See Justice *et al.* 1985). Between about 1.3 and 2.7 μm, absorption of infrared radiation by water molecules decreases the albedo. (This absorption is also responsible for the small increase in the albedo of pure water at these wavelengths.)

The abrupt and characteristic change in the reflectance of vegetation at about 0.7 μm is the basis of a number of techniques for assessing the amount of vegetation present in a remotely-sensed image (Harris 1987). In the simplest type of application, a *vegetation index* is calculated as

$$\frac{I_2 - I_1}{I_2 + I_1}$$

Fig. 3.7. Typical spectral albedos (schematic) of various materials in the visible and near infrared bands.

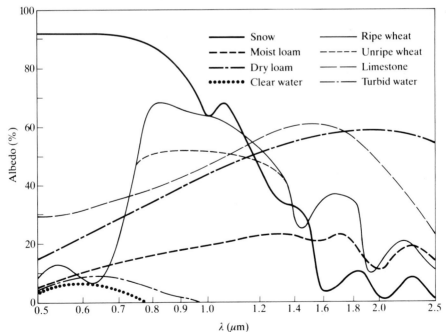

where I_1 is the intensity of the radiation reaching the sensor from a given area in the visible part of the spectrum, and I_2 is the intensity in the near infrared part of the spectrum. This index is then assumed to bear a monotonic relationship to the fraction of the area which is vegetated.

The high and comparatively uniform reflectance of *snow* in the visible and near infrared bands is caused by the large number of nonselective (i.e. wavelength-independent – see section 3.4.2.1) scattering events which a photon undergoes at the many interfaces between ice and air. A similar argument explains the white colour of *clouds*. The high reflectance of snow means that under favourable conditions it can be detected by a satellite-borne sensor when illuminated only by *moonlight* (Foster 1983).

The spectral reflectance properties of *rock* are less distinctive than those of vegetable matter, in the sense that it is much harder to recognise a single spectral type that is 'rock-like' than one which is 'vegetation-like'. Mineral reflectance spectra are strongly dependent on the chemical composition of the rock (see e.g. Elachi 1987). For example, iron oxides give characteristic minima in reflectance below $0.5\,\mu m$ and at about $0.9\,\mu m$, whereas carbonates and clays are dominated by a number of narrow absorption features between 2.1 and $2.5\,\mu m$. Weathering of rock, and the local roughness properties of its surface, also play a part in modulating the spectral reflectance.

The reflectance properties of *soil* are clearly dependent on the type of rock from which the soil is derived, but they are also strongly dependent on the content of vegetable matter and water, both of which tend to lower the albedo (especially, in the case of water, in the infrared). There is also a possible interaction between soil type and the spectral reflectance of vegetation growing in the soil, since trace elements in the soil may influence the growth pattern of the vegetation. The practical application of this effect is *geobotany*, in which geological prospecting is carried out by observing changes in the colour of vegetation. It is particularly useful for the detection of copper.

In the visible and near infrared regions of the electromagnetic spectrum we will generally be concerned with broad-band illumination from the sun or sky, but there is no reason why we should restrict ourselves in this way. We could imagine a remote sensing instrument which both emitted and received radiation over narrow spectral ranges, and in this case we would normally expect that the scattered radiation would lie in the same part of the spectrum as the incident radiation. This is not necessarily so, however. We have already seen (in chapter 2) one mechanism whereby the

Table 3.2. *Typical values of infrared emissivity (10 μm)*

Material	Emissivity
metals	0.01–0.6
snow (compressed)	0.7–0.85
ice (glacier)	0.85
soil (dry sandy)	0.88–0.94
wood	≈0.9
granite (rough)	0.90
plaster	0.91
concrete	0.92
soil (dry loamy)	0.92
brick	0.93
glass	0.94
soil (moist)	0.94–0.95
sand	0.95–0.96
leaves (dry)	0.96
road (tarred)	≈0.97
snow	0.97–1.00
ice	0.98
skin (human)	0.98
peat	0.98
grass (green)	0.98–0.99
leaves (damp)	0.99
water	0.99

scattered and incident radiation have different spectra, namely solar radiation heating a surface and so giving rise to thermal emission. However, there is another important mechanism which changes the wavelength, and this is *fluorescence*. This is a quantum-mechanical effect which has no intermediate thermal stage. Many minerals show fluorescence (for example fluorspar, from which the effect derives its name), but chlorophyll also displays this phenomenon, fluorescing in the wavelength range 0.7 to 0.85 μm. This effect is small, and not normally noticeable, but it can be detected (for example) at night in chlorophyll-containing marine organisms, when stimulated by laser radiation of shorter wavelength. It can also be detected by sensitive passive visible-wavelength sensors. The absorption length of water in this range of wavelengths is about 3 m, hence this is the maximum depth at which the concentration of such organisms can be detected.

There is of course no need to confine our consideration of spectral albedo to the visible and near infrared bands. Table 3.2 lists the emissivities of various materials in the thermal infrared region, and fig.

3.8 shows the approximate variation of emissivity of a few materials in the microwave region, from 1 to 36 GHz. A discussion of the processes which modulate these curves is beyond the scope of this book (see e.g. Long, 1983; Ulaby *et al.*, 1986), but again we see the advantage of a multifrequency approach. It is particularly important to be able to make multifrequency measurements in passive microwave radiometry, because the poor spatial resolution inherent in a system operating at long wavelengths (2.32), especially when observing from a satellite, means that a single resolution element is likely to contain a variety of materials. Multispectral or dual polarisation techniques offer a possible means of identifying these materials. Let us suppose for example that we are observing a polar ocean at a certain frequency, and that we know that the signals which would be measured from pure water, old ice and young ice are T_w, T_o and T_y respectively at that frequency. Since the area under observation, because of its large size, probably contains a mixture of ice and water, the observed signal T_b will be a weighted average of these values. If f_o and f_y are the fractions of the area occupied by old and young ice respectively, we may put

$$T_b = T_w(1 - f_o - f_y) + T_o f_o + T_y f_y$$

Fig. 3.8. Microwave emissivities in the normal direction (schematic) of various materials.

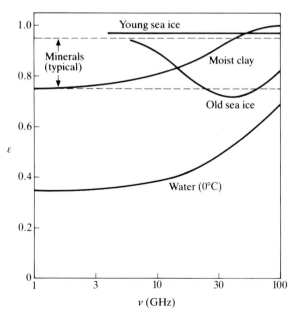

ν (GHz)

It is clear that this equation is insufficient to determine the two unknown parameters f_o and f_y. However, if we make an observation of the same region at a different frequency, we may derive a second equation of the same form, allowing the two unknown parameters to be determined. This is of course merely a specific statement of a general principle: that we must make N independent observations in order to determine N independent parameters.

3.3 Penetration of the target

In section 3.2 we discussed the way in which radiation is reflected from a surface, and we have also noted the connexion between the emissivity of a surface and the total power reflected from it. We must now consider what happens if a significant fraction of the incident radiation penetrates below the surface. If the medium below the surface is homogeneous, the radiation is transmitted through it, although it may also be absorbed. We saw in section 2.2 that, in general, when an electromagnetic wave propagates into a homogeneous absorbing medium its electric field strength may be written

$$E = E_0 \exp[i\omega(t - n'z/c)] \exp(-\omega\kappa z/c) \tag{2.18}$$

where n' and $-\kappa$ are the real and imaginary parts of the refractive index, and z is the distance propagated into the medium. We saw also in section 2.2 that this represents a wave whose intensity falls by a factor of e^2 (about 7.4) each time it travels a further distance

$$l_a = c/\omega\kappa \tag{2.19}$$

where l_a is called the absorption length. Fig. 3.9 shows the variation with wavelength of the absorption lengths of various materials.

Roughly speaking, we may say that the radiation which penetrates the surface is absorbed within a distance of the order of l_a, unless of course the material does not extend this far in which case we need to consider also the effect of whatever is below this layer. This would be the case, for example, if we were to consider the visible-wavelength radiation incident upon a body of clear water only a few metres deep. Since the albedo of water is only 0.02 most of the incident radiation enters the bulk of the liquid; and since the absorption length is many metres, most of this radiation will in fact reach the bottom. Thus unless the albedo of the bottom surface is also very low, the signal detected by a sensor will be dominated by the contribution from the bottom surface.

3.3.1 Volume scattering

The general theory of the transfer of radiation through a scattering medium is well beyond the scope of the present book, although we shall say something more about scattering in section 3.4, so here we shall merely make a few remarks about volume scattering.

If the target material is not homogeneous, any radiation which penetrates the surface will be scattered to some extent by the inhomogeneities, as well as being absorbed. The combined effects of scattering and absorption in reducing the intensity of the forward-going wave are collectively referred to as *attenuation*, and an *attenuation length* is defined analogously to the absorption length.

The first important point to note about volume scattering is that, since some radiation is scattered back out of the target material, it is an effect which adds to the surface scattering effects discussed in section 3.2. In an analogous way to the theory developed in that section, it is clear that the

Fig. 3.9. Absorption lengths (schematic) of various materials. Note that the absorption lengths are strongly influenced by such factors as temperature and the content of trace impurities, especially at low frequencies.

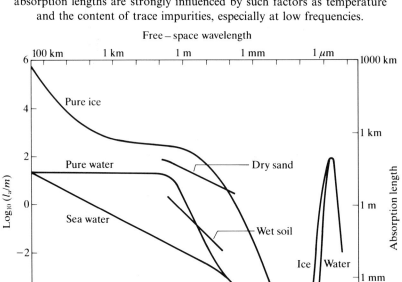

volume scattering will depend on both the geometrical and dielectric properties of the medium, only now it must be the three-dimensional geometrical properties which need to be considered. The reflected signal will in general be composed of a part which has been scattered from the surface, and a part which has been scattered from within roughly one attenuation length of the surface, and separating these two contributions may well present difficulties. On the other hand, the scattered radiation clearly contains information about the internal structure of the target, and if there is an adequate theory of the surface and volume scattering, important structural parameters may be derivable.

Volume scattering effects are of the greatest importance in active microwave remote sensing, where significant penetration can sometimes occur, as for example into a dry snowpack or a vegetation canopy (see e.g. Tsang *et al.*, 1985; Ulaby *et al.*, 1981, 1982, 1986).

3.4 Interaction of electromagnetic radiation with the atmosphere

So far in this chapter we have discussed the interaction of electromagnetic radiation with the surface and bulk of the material being sensed. However, the radiation also has to make at least one journey through at least part of the earth's atmosphere, and two such journeys in the case of systems which detect reflected radiation, whether artificial or naturally occurring. Each time a ray passes through the atmosphere, it undergoes both absorption and scattering. Again, the sum of these two forms of energy loss is called the *attenuation*. In addition, the atmosphere has a refractive index which differs from unity, so that radiation travels through it at a speed different from the vacuum speed of $299\,792\,458\,\mathrm{ms}^{-1}$. This is potentially important for any system (such as those discussed in chapter 7) which uses the reflexion of electromagnetic pulses to determine range. All of these effects are, in general, dependent on wavelength.

3.4.1 Composition of the atmosphere

In order to understand the effect of the atmosphere on electromagnetic radiation, we must first know something about its structure and composition. At sea level, the principal constituents of the dry atmosphere are nitrogen (78.08% by volume), oxygen (20.95%), argon (0.93%) and carbon dioxide (0.03%). There are also traces, measured in parts per million, of neon, helium, krypton, xenon, hydrogen, methane and nitrous oxide, and other even rarer species. (These figures refer to the gases composing the dry part of the air.) There is also a significant but variable amount of water vapour. The proportion of water vapour can be calculated if the air temperature T and the relative humidity RH (expressed as a fraction) are known. The partial pressure of water vapour is

$$p_{water} = p_v(T) \cdot RH \qquad (3.7)$$

where $p_v(T)$ is the saturated vapour pressure of water at the temperature T (see appendix). If the total air pressure is p_0, the volume fraction of water vapour is then p_{water}/p_0. Thus for example if the air temperature is $20\,°C$, $p_v(T) = 2.34 \times 10^3\,Pa$ and so, if the humidity is $80\% = 0.8$, the partial pressure of water vapour is $1.87 \times 10^3\,Pa$. Since the normal sea-level atmospheric pressure is $1.01 \times 10^5\,Pa$, the volume fraction of water vapour is about 1.8%. In addition to these gaseous constituents, the atmosphere contains liquid and solid water (in clouds and in the form of precipitation), dust, and aerosol particles. The concentration of these materials, like that of water vapour, is variable.

Atmospheric pressure and density diminish with height above the earth's surface. This is because the molecules, acted upon by gravity, attempt to sink to the surface but are prevented from doing so by thermal motion. The distribution of density with height is thus governed by the Boltzmann distribution, and so is roughly exponential. There are, however, significant variations from this approximate dependence, and it is conventional to divide the atmosphere into several layers. These are the *troposphere* (approximately 0–11 km above the earth's surface), in which the temperature decreases with height, the *stratosphere* (11–50 km), in which the temperature is approximately constant up to 35 km and then increases with height, the *mesosphere* (50–80 km), and the *thermosphere* (above about 80 km). Fig. 9.11 shows the variation of pressure, density and temperature with height in the standard atmosphere, from which it can be seen that 90 per cent of the atmospheric mass is below a height of about 16 km. Thus a satellite remote sensing system looks through effectively all of the atmosphere, whereas an airborne system looks through only part of it.

3.4.2 Atmospheric attenuation

As electromagnetic radiation propagates through the atmosphere, it will interact with the various species it encounters. This interaction may, as we have seen, take the form of *scattering*, where energy is not lost but merely redirected, or of *absorption*, where energy is absorbed by the species (and, to conserve energy, ultimately reradiated although again in different directions and probably over a different range of wavelengths). In either case, energy is lost from the forward direction, and the combined effect of scattering and absorption is called *attenuation*.

We saw in chapter 2 that, as a result of absorption in a homogeneous medium, the intensity of an electromagnetic wave decreases exponentially with distance. In an inhomogeneous medium of uniform density the same is true of attenuation by both absorption and scattering (at least, this is true unless the absorption is caused by spectral lines), and so it is convenient to define the extent of the attenuation on a logarithmic scale. This is done using the *decibel* as a (dimensionless) unit, defined by (3.8).

$$A = 10 \log_{10}(I_1/I_2) \tag{3.8}$$

In this equation, A is the attenuation in dB (decibels), I_1 is the incident (unattenuated) intensity and I_2 is the output (attenuated) intensity. It can be seen from this equation that an increase in A of 10 dB results in a decrease in I_2 by a factor of 10; and an increase in A of 20 dB implies a decrease in I_2 by a factor of 100 and so on.

A related parameter is the *optical depth* τ. This expresses the extent to which a given layer of material attenuates the intensity of the radiation passing through it. The optical depth (sometimes called the *opacity*) is defined as

$$\tau = \ln(I_1/I_2) \tag{3.9}$$

Thus,

$$\tau = 0.2303 \, A \tag{3.10}$$

If τ is much greater than 1 for a given sample of material, the sample is said to be *optically thick* (or opaque), and conversely if τ is much less than 1 the sample is optically thin (transparent). The total optical depth of the atmosphere at a wavelength of $0.55 \, \mu m$ is about 0.3 under normal conditions, 0.1 being contributed by atmospheric gases and 0.2 by particulate matter.

Recalling (2.16), we may, for a wave travelling in a uniform medium of constant density, put

$$I_2 = I_1 \exp(-2z/l_a) \tag{3.11}$$

(the *Lambert–Bouguer law*) where l_a is now the amplitude attenuation length for both scattering and absorption combined, and z is the propagation distance. Comparing (3.8) and (3.11) we find that

$$A = 20(\log_{10}e)z/l_a$$

Thus the attenuation in dB is proportional to the distance travelled through the medium, and we can define an *attenuation coefficient* α measured in decibels per unit length. It is clear from the above equation that

$$\alpha = 20(\log_{10}e)/l_a = 8.69/l_a \qquad (3.12)$$

Equation (3.11) is valid only if the density of the medium is constant, and of course the density of the atmosphere decreases with height. However, it is clearly reasonable to use (3.11) for the lowest kilometre or so of the atmosphere; and at satellite altitudes, above essentially all of the earth's atmosphere, we may instead quote the total atmospheric attenuation. Fig. 3.10 shows the total atmospheric attenuation for a vertical ray. The figure is somewhat schematic, especially in the regions affected by water vapour which is of course a variable constituent. Elaborate models of the attenuation of the 'standard atmosphere' are available, some of these in a form which can be used by a computer.

Let us try to understand in detail the shape of fig. 3.10.

Fig. 3.10. Total zenith attenuation by the atmosphere under normal conditions (simplified).

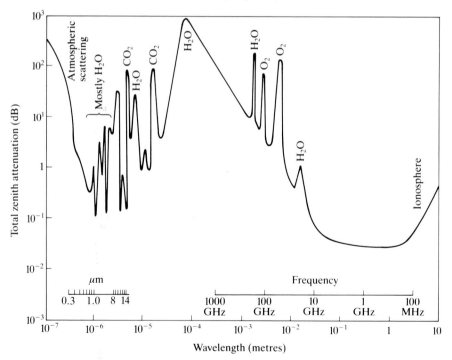

3.4.2.1 Rayleigh scattering

Below about $0.4\,\mu m$ (i.e. in the blue to ultraviolet region of the spectrum) the attenuation is dominated by scattering from individual molecules. When the scattering particles are small compared with the wavelength, as is clearly the case, the kind of scattering which occurs is known as Rayleigh scattering.

The scattering properties of particles are often specified in terms of the *scattering cross-section* σ, which has the units of an area and is defined such that if a flux density $F\,(\mathrm{Wm}^{-2})$ is incident on a single particle, the scattered power is $F\sigma$. If we consider radiation of flux density $F\,(\mathrm{Wm}^{-2})$ vertically incident on a 'slab' of atmosphere of area A and height (thickness) dz, the number of particles encountered by it is $nAdz$, where n is the number density of the particles. The scattered power is thus $nA\sigma Fdz$, and we can thus write

$$\frac{dF}{F} = -n\sigma dz$$

as the differential equation governing the attenuation of the flux density. Integrating this expression through the whole atmosphere gives

$$\tau = \sigma \int_0^{\infty} n\,dz$$

$$= \sigma N_t \tag{3.13}$$

N_t is the integrated number density, and can clearly be interpreted as the total number of molecules (or whatever) in a column of area $1\,\mathrm{m}^2$ extending indefinitely through the atmosphere. It may easily be shown that

$$N_t = p_0/mg \tag{3.14}$$

where p_0 is the sea-level partial pressure of a molecular species, and m is the mass of the molecule. Finally, we need to borrow a result from classical electrodynamics. If we model our scattering species as conducting spheres of radius a (which is a reasonable model for molecules), the Rayleigh scattering cross-section can be shown (e.g. Jackson 1975) to be

$$\sigma = 128\pi^5 a^6/3\lambda^4 \tag{3.15}$$

where λ is the wavelength. (Note that for a dilute gas, with refractive index $n' \approx 1$, a more general formulation for the scattering cross-section is

$$\sigma = 32\pi^3(n'-1)^2/3\lambda^4 n^2 \tag{3.16}$$

where n is again the number density of molecules.)

The proportionality of σ to λ^{-4} shows that blue light is very much more strongly scattered than red light. Thus a beam of 'white' light from the sun, travelling through a column of air, will suffer selective scattering in the sense that much of the blue light will be removed from the forward direction and redistributed sideways. This is why the sky appears blue, and why the rising or setting sun appears red even in the absence of scattering by dust particles. We should also note that the Rayleigh scattering process will in general increase the degree of polarisation of the scattered radiation.

Let us estimate the radius of an 'air molecule' as 1.15×10^{-10} m, half way between the oxygen–oxygen and the nitrogen–nitrogen bond lengths. Substitution into (3.15) gives $\sigma = 1.2 \times 10^{-30}$ m^2 at $0.4\,\mu$m and 3.0×10^{-28} m^2 at $0.1\,\mu$m. Taking $p_0 = 1.0 \times 10^5$ Pa, and $m = 4.8 \times 10^{-26}$ kg (for an 'air molecule'), we obtain $N_t = 2.1 \times 10^{29}$ m^{-2}, and substituting this into (3.13) and (3.10) we find that $A = 1$ dB at $0.4\,\mu$m and about 300 dB at $0.1\,\mu$m, roughly the values shown on fig. 3.10.

Other kinds of atmospheric particle, as well as 'air molecules', can scatter radiation. In general, if the particle is very much smaller than the wavelength of the radiation, Rayleigh scattering dominates, and the scattering cross-section is proportional to λ^{-4} as we saw in (3.15). If the particle is very much *larger* than the wavelength, the scattering is said to be *non-selective*, and the cross-section is independent of the wavelength. The treatment of intermediate cases, usually referred to as *Mie scattering*, is complicated (see e.g. van de Hulst 1981 for details), but may often be crudely approximated as being proportional to λ^{-1}.

The smallest non-molecular particles which are important for atmospheric scattering are aerosols (dispersed systems of small particles suspended in a gas), of which haze is an example. These particles are caused by surface and cosmic dust, by volcanic ash, combustion products, and salt crystals (around which water molecules cluster). The typical particle size is 0.01–$1\,\mu$m, and the typical number density 10^7–10^9 m^{-3}. Next in size come fog particles, which are typically 1–$10\,\mu$m and have a density of 10^7–10^8 m^{-3}, followed by cloud and rain (1–$10\,\mu$m and 10^2–$10^4\,\mu$m in size, and 10^7–10^9 and 10^3–10^4 m^{-3} in density, respectively).

3.4.2.2 *Molecular absorption*

Molecules have three main mechanisms by which to absorb energy from an incident electromagnetic wave. In this section we can give only a brief sketch of what is a large subject. Schanda (1986) gives a more detailed treatment of molecular absorption as applied to remote sensing.

The first mechanism, requiring the most energy, involves the promotion of electrons to higher energy levels. These are termed *electronic transitions*, and the calculation of the energies involved is beyond our scope. They are of the order of electron volts, so that the corresponding wavelengths are of the order of 1 μm. Of the remaining two mechanisms for the absorption of energy, the first is to put it into the form of vibration of the atoms, and the second is to put it into rotational kinetic energy. We may estimate the energy, and hence the frequency, of these types of absorption as follows.

The molecular bond between atoms behaves more or less as a spring, so we shall denote the force constant of such a spring by $k\,(\text{Nm}^{-1})$. For a diatomic molecule consisting of atoms of mass m_1 and m_2, the vibration frequency is given by

$$v = \frac{1}{2\pi}\left(\frac{k(m_1+m_2)}{m_1 m_2}\right)^{\frac{1}{2}} \tag{3.17}$$

Quantum mechanics tells us that (to first order) such a system will absorb radiation only at the frequency v, i.e. that it will generate a spectral absorption line. In order to estimate a typical value for v from (3.17) we need to know a typical value of the molecular spring constant k. This can be done very roughly by realising that since the elastic properties of solids are also governed by the stiffness of the interatomic bonds, k can be estimated from a typical value of the elastic modulus E. In fact, if a is a typical interatomic separation, we expect that $E \approx k/a$. Putting $a = 10^{-10}$ m and $E = 5 \times 10^{11}\,\text{Nm}^{-2}$ (a reasonable value for very stiff materials), we see that typically $k = 50\,\text{Nm}^{-1}$. This is expected to be an underestimate of the spring constant in a diatomic molecule, since in a solid there will probably be significant delocalisation of the electrons forming the bonds. Thus we might guess that, in a diatomic molecule, $k \approx 500\,\text{Nm}^{-1}$. Putting $m_1 = 2.0 \times 10^{-26}$ kg and $m_2 = 2.7 \times 10^{-26}$ kg to represent a molecule of CO, and substituting into (3.17), we find that a typical value of v is 3.3×10^{13} Hz ($\lambda = 9\,\mu$m).

Naturally, the estimate given above for the spring constant k of a molecular bond is crude, and serves merely to show that we may expect to find molecular vibrational absorption lines in the infrared region of the spectrum. In fact, the vibration frequency of the CO molecule is 6.5×10^{13} Hz, so our estimate is a reasonable one, and infrared absorp-

Table 3.3. *Important molecular absorption lines in the atmosphere*

Molecule	Visible and infrared bands Wavelengths (μm)
H_2O	0.9, 1.1, 1.4, 1.9, 2.7, ≈ 6
O_2	0.8
CO_2	2.7, 4.3, ≈ 14
N_2O	4.6, 7.7
O_3	9.5
Molecule	Microwave band Frequency (GHz)
H_2O	22.235, 183.3
O_2	≈ 60, 118.75

tion features are found at wavelengths down to about 1 μm. Table 3.3 lists the most important infrared absorption lines.

The other important mode of molecular absorption is rotation. If a molecule has a moment of inertia I, quantum mechanics shows that its kinetic energy of rotation takes on values of

$$E_r = J(J+1)h^2/8\pi^2 I$$

where h is Planck's constant and J is a quantum number which can take only integral values (0,1,2, . . .). When the molecule absorbs radiation, the quantum number increases by one so that the energy absorbed is $(J+1)h^2/4\pi^2 I$. Using Planck's law $E=hv$ to relate the energy to the frequency, we find that the frequency of the absorption line is given by

$$v = (J+1)h/4\pi^2 I \tag{3.18}$$

The moment of inertia I of a diatomic molecule is

$$m_1 m_2 a^2/(m_1 + m_2)$$

where a is the interatomic distance. Again using our values for the CO molecule as an example, and taking 1.13×10^{-10} m for a, we find that $I = 1.5 \times 10^{-46}$ kgm^2, and that v takes on integral multiples of 1.1×10^{11} Hz. Thus we expect to find molecular rotational absorption lines at multiples of 100 GHz or so, and at lower frequencies for larger, heavier molecules. The important molecular rotations thus occur in the microwave region of the spectrum, and are also shown in table 3.3.

3.4.2.3 The ionosphere

The ionosphere is an ionised layer above the earth's atmosphere, extending from about 70 km to a few hundred km above the earth's surface. The ionisation is produced by extreme ultraviolet and X-radiation from the sun. If n_e is the number density of electrons in the ionosphere, the refractive index may be shown to be

$$n = [1 - n_e e^2 / m\omega^2 \varepsilon_0]^{\frac{1}{2}}$$

as we saw in chapter 2. As long as $\omega \gg \omega_p$, n is real and the refractive index may be written as approximately

$$n \approx 1 - \omega_p^2 / 2\omega^2$$

where ω_p is the *plasma frequency*, given by

$$\omega_p^2 = n_e e^2 / \varepsilon_0 m \qquad (3.19)$$

We see however that, if $\omega < \omega_p$, the refractive index becomes purely imaginary and the wave will be strongly attenuated. Since the maximum electron density in the ionosphere is of the order of 10^{12} m^{-3}, ω_p is about 5×10^7 s^{-1} or 9 MHz. (The ionosphere is, however, notoriously variable both spatially and temporally, with variations in electron density being caused by changes in geomagnetic latitude, time of day (especially near dusk and dawn) and year, the number of sunspots on the face of the sun, and so on.) Thus for frequencies below about 9 MHz the ionosphere is opaque, and this sets a lower frequency limit for satellite-borne remote sensing. As the frequency increases above this value, the ionosphere becomes progressively more transparent, as shown in fig. 3.10. (It is worth noting here that the opacity of the ionosphere to frequencies below a few MHz is beneficial, in that it allows HF (high-frequency or short-wave) radio signals to propagate for long distances round the earth's surface. The signals are reflected at the earth's surface and at the lower surface of the ionosphere.)

We have thus noted the main features of the graph of atmospheric attenuation shown in fig. 3.10. There are two practical consequences of this variation with frequency. The first is that there exists a number of *spectral windows*, regions of the electromagnetic spectrum in which the earth's atmosphere is effectively transparent, and in which observation of the surface from the air or from space is feasible. Table 3.4 lists the most important of these spectral windows.

The other important point to note about atmospheric attenuation is that, where it is high, the atmosphere must *emit* radiation as a result of thermal excitation. This follows from the same argument, due to Kirchhoff, which we employed in section 2.6 to show that a body which is a good absorber of radiation must also have a high black-body emissivity. In a uniform material the bulk of the detected radiation will arise from

Table 3.4. *Important atmospheric spectral windows*

Visible and infrared bands (wavelengths in μm)	
0.3–0.9	2.0–2.4
1.0–1.1	3.5–4.0
1.2–1.3	4.6–4.9
1.5–1.8	8.0–13.0 (some absorption lines)
Radio band (frequencies in GHz)	
0.1–15	140–160
25–35	230–250
80–100	260–290

within about one attenuation length of the sensor, although if the density of the absorbing species decreases with height, as is normally the case, the two effects combine to result in most of the radiation being received from a distinct layer of the atmosphere. This is the basis of remote sensing techniques used for sounding the atmosphere; by tuning the observation frequency, the height of this layer may be varied, and parameters such as atmospheric temperature and pressure may then be determined as a function of height. However, as was mentioned in the preface, this is beyond the scope of the present work and we must leave the subject of atmospheric sounding at this point. The interested reader will find details in, for example, Chen (1985) or Elachi (1987).

3.4.3 *Atmospheric and ionospheric turbulence*

One further important influence which the atmosphere has on electro-magnetic radiation passing through it is that of atmospheric turbulence. This is always present to a greater or lesser extent in the lower atmosphere, and causes variations in the density, and hence the refractive index, of the air. The phase of an electromagnetic signal is corrupted by these variations, and this adversely influences the behaviour of an imaging system.

   The most useful way to describe the effects of this kind of phase fluctuation, which is of course statistical rather than deterministic, is by specifying the *structure function*. This is usually defined as the variance of the phase difference between two points which would, in the absence of such effects, be in phase. This function is therefore measured in radians squared, and is usually a function of wavelength, and of the separation of the two points. The timescale over which the phase variance is measured is also often important.

So far as the effect on the resolution of an imaging system is concerned, an approximate rule of thumb is to replace the turbulent medium by a notional aperture whose size is equal to the separation at which the structure function attains a value of 1 radian squared. For light passing through the whole of the earth's atmosphere, this separation is about 0.2 metres, corresponding to an angular resolution of about 3×10^{-6} radians or about 1 second of arc. This is a significant limitation to the ground-based, upward-looking observations of terrestrial astronomers. However, for the downward-looking observations with which we are concerned, what is of greater concern is the *linear* resolution at the surface. The limit imposed in this case (see fig. 3.11 for an illustration of this point) will be of the order of 3×10^{-6} radians multiplied by the scale height of the atmosphere, giving a linear resolution limit of about 0.02 m. This is unlikely to be a serious problem to remote sensing in the immediate future.

Fig. 3.11. Scattering caused by atmospheric turbulence (very schematic). A ray is scattered into a cone of angle θ_s. (a) shows the effect this has on an astronomical observation. Radiation from a point source at the zenith can be received at P from all directions between A and B, so that the source appears to subtend the angle θ_s. (b) shows the effect of such scattering on a remote sensing observation. Radiation can emerge from the atmosphere at the point Q travelling towards the zenith having been emitted from any point between A and B. A point object P thus appears to be spread out over a horizontal distance $H\theta_s$, where H is the effective height of the atmosphere.

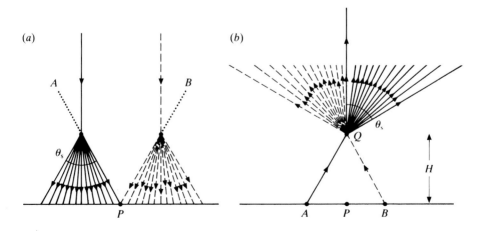

Turbulence in the lower atmosphere causes a similar effect at radio frequencies, because of the similar refractive index of air at radio and optical wavelengths. However, for observations from satellites at radio frequencies, the *ionosphere* provides a potentially more serious problem. It is not really possible to quote a typical value for the ionospheric structure function, because of the great variability alluded to earlier, but we may note that the phase variance will be proportional to λ^2 (because of the plasma dispersion relation, discussed in chapter 2), and that it will be greater near the geomagnetic poles and the geomagnetic equator.

3.4.4 *Cloud, rain and snow*

At any one time, about half of the earth's surface will be covered by cloud. Visible-wavelength sensors, and to some extent those operating in other wavebands of the electromagnetic spectrum, will be limited by the presence of significant amounts of cloud cover. This can be a serious problem in the case of satellite-borne sensors which revisit a particular location comparatively rarely (see chapter 9). For example, it has been estimated that the LANDSAT satellite, which revisits each location every 16 days, will obtain a cloud-free scene of a certain location in the UK only once per year, and a scene with 1 okta of cloud cover (1 okta is one eighth of the sky obscured by cloud) only twice per year. Similarly, of the 200 000 scenes acquired by the SPOT-1 satellite by 1987, only 24% had less than 1.6 okta of cloud. The probability of less than 10% cloud cover in a single LANDSAT observation over the continental USA is about 0.05–0.4, and the number of observations needed to obtain a probability of 0.75 of less than 10% cloud cover is between 5 and 100 (Goetz 1979).

The influence of precipitation on the propagation of electromagnetic radiation through the atmosphere is also significant. Broadly speaking, this influence may be characterised as an attenuation coefficient dependent on the rate of precipitation and the observing frequency, and chapter 6 gives details. In the microwave region, volume scattering from rain can usefully be applied as a sounding technique to determine the rainfall rate.

Problems

1. Explain the meanings of the terms radiance, irradiance and radiant exitance, and define the bidirectional reflectance distribution function (BRDF) of a surface in terms of these quantities. Describe

briefly the principal methods by which the BRDFs of rough surfaces may be calculated.

A given (rough) surface has a BRDF which is proportional to $\cos(\theta_0)\cos(\theta_1)$, where θ_0 and θ_1 are the angles of the incident and scattered radiation directions, measured from the normal to the surface. The BRDF has no azimuthal dependence. Show that, if the albedo for normally incident radiation is 1, the diffuse albedo of the surface is 2/3.

2. Use the facet model to derive the angular distribution of power reflected from a one-dimensional 'surface' consisting of facets (lines) of length D and slope β. β has a Gaussian distribution with zero mean and standard deviation β_0. Assume that $D \gg \lambda$ and $\beta_0 \ll 1$, and show that the scattering is dominated by β_0 if the rms variation in surface height is greater than about $\lambda/4$.

3. Show that the relative permittivity of a plasma is given by

$$\varepsilon = 1 - Ne^2/\varepsilon_0 m\omega^2$$

where N is the electron density, m is the electron mass and ω is the frequency. Explain why the ionosphere presents a barrier to electromagnetic waves below a certain frequency.

A dual-frequency satellite radar altimeter simultaneously emits short pulses at 2.00 GHz and at 5.00 GHz. The reflected pulses are found to be separated by a time interval of 15.0 ns. Calculate the ionospheric total electron content (TEC).

4. Show by comparison of (3.16) and (2.16) that an electron has a scattering cross-section of $\mu_0^2 e^4/6\pi m^2$. (This is known as the *Thomson cross-section.*) Calculate the effective scattering radius r_e of the electron, if this cross-section is given by $8\pi r_e^2/3$.

5. Explain why the limit to the angular resolution of a satellite-borne remote sensing system set by atmospheric turbulence is smaller than that for an earth-based astronomical system.

4

Photographic systems

4.1 Introduction

Aerial photography, as we remarked in chapter 1, represents the earliest modern form of remote sensing system. It is thus the most familiar system, and represents the best point at which to begin our discussion of types of imaging system. As we have seen in chapter 1, remote sensing systems may conveniently be classified as passive or active, imaging or non-imaging, and aerial photography is evidently a passive imaging system. We may also characterise it by saying that it is not a scanning system, but we shall reserve our discussion of the distinction between scanning and non-scanning systems until the next chapter.

In this chapter we shall consider the construction, function and performance of photographic film, and in particular its use for obtaining quantitative information about the size of objects. The chapter then discusses the effects of atmospheric scattering on aerial photography, and finally gives a brief account of the applications of the technique. Further details on the physical and chemical aspects of the photographic process can be found in Engel (1968) and in Open University (1978), and of course there are very many books on aerial photography and photogrammetry (for example American Society of Photogrammetry 1981) to which the reader may refer.

4.2 Photographic film

Fig. 4.1 shows schematically the construction of photographic film. An 'emulsion' (it is not strictly an emulsion) consisting of grains (crystals) of silver bromide, silver chloride or silver iodide held in suspension in a layer of gelatin about 100 μm thick is supported on a plastic base. The base material merely serves to provide a mechanical support for the emulsion. The grains in the emulsion are typically 1 to 10 μm in size.

As we shall see in this and subsequent chapters, the fundamental principle involved in detecting electromagnetic radiation is to observe changes in the energy of electrons. Since the electron is a charged particle of low mass, it is easily acted upon by the electric field of the radiation. In the case of photographic film, the interaction which takes place is that an electron is transferred from a bromide (or other halide) ion to a silver ion, leaving an atom of silver and an atom of bromine:

$$Ag^+Br^- \rightarrow Ag + \tfrac{1}{2}Br_2$$

Normally, this reaction reverses itself spontaneously, but if a sufficient number of silver atoms (typically about four) is formed, in a short space of time, in a single grain, the clump of atoms is stable. This clump is then known as a *development centre*, and the film is said to contain a *latent image*. After its exposure to light, the film is *developed* by treating it with a chemical reducing agent. This is carefully chosen so that it is capable of reducing Ag^+ to Ag only in the presence of existing silver atoms, which act as nuclei for the growing silver crystals. Thus, only those grains which contain development centres are converted to solid silver. The remainder are unchanged, and are later washed away.

We can see from this brief description of the theory (the Gurney–Mott theory) of the photographic process that an exposure to light sufficient to cause only a few atoms of silver to be formed in a given grain finally converts the entire grain, containing perhaps 10^{10} atoms, to silver. There is thus an amplification by a factor of about 10^9 involved in the process of

Fig. 4.1. Schematic diagram showing the construction of a typical photographic film.

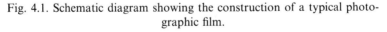

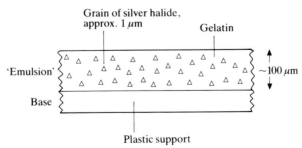

photographic detection. A more complete account of the Gurney–Mott theory is provided by Omar (1975).

4.2.1 Film types

4.2.1.1 Black and white film

This is the simplest type of film, and that used in the earliest aerial photography. It is still in wide use.

The photochemical reaction on which the photographic process depends has already been discussed. Since it depends on the oxidation of the halide ion, the energy required for this process will govern the maximum wavelength (minimum photon energy) to which the film can respond. For the halide ions in the crystalline state, these maximum wavelengths lie in the range 0.4–0.5 μm. Such films ought therefore to respond only to blue, violet and UV radiation (and in fact to shorter wavelengths, as far as X-rays. It is the opacity to these wavelengths of the optical glass and the gelatin of the film which prevents their being problematical). In fact, the range of sensitivity is extended in many films by the use of *sensitising dyes*. Panchromatic (i.e. 'all colours') film has a sensitivity range from about 0.35 to 0.65 μm, and infrared-sensitive films can be made sensitive to wavelengths as long as 1.2 μm.

By the use of suitable films and filters, photographs can be made to respond only to the range (say) 0.7–0.9 μm, in the near infrared. This region of the electromagnetic spectrum gives improved penetration of haze, since longer wavelengths are less affected by scattering (see sections 3.4.2.1 and 4.6). Near infrared radiation is strongly reflected by vegetation, and strongly absorbed by water (see chapter 5), so studies of either of these materials are often aided by near infrared photography.

We have noted that black and white film is intrinsically sensitive to ultraviolet radiation (0.3–0.4 μm). These wavelengths are normally filtered out, because of the reduction in contrast caused by the strong scattering. However, it is occasionally useful to record *only* the UV wavelengths, for example when observing fluorescent materials such as carbonate minerals and oil films.

4.2.1.2 Colour film

The human eye can distinguish only about 200 shades of grey, but about 20 000 tints and shades of colour. Thus far more information can be deduced from a colour image than from a black and white image, even if the means of doing so is merely visual examination. (Ray & Fischer 1957, Yost & Wenderoth 1967.)

Colour films are constructed with three layers of emulsion instead of just one. By the incorporation of suitable dyes, the uppermost layer is made sensitive to blue light, the middle one to blue and green, and the bottom one to blue and red. A yellow filter is interposed between the top and middle layers to prevent blue light from exposing all three layers.

Because of the complexity of its construction and, in consequence, processing, colour photography is expensive. Nevertheless, airborne colour aerial photography has found a wide variety of applications, especially in agriculture, forestry, ecology, geomorphology, hydrology and oceanography. Plate 1 shows an example of spaceborne colour photography.

As well as 'conventional' colour film, *false colour infrared* (FCIR) film is also available. This was originally developed during the Second World War to assist in the location of enemy tanks which were using woods for camouflage, and it has subsequently found many uses in civilian remote sensing. These uses have mainly been in the study of vegetation, but FCIR film has also been used in studying urban areas, mapping wetlands (see plate 2), finding archaeological crop marks and so on. The construction of the film is similar to that of conventional colour film, but the layers of emulsion respond not to blue, green and red (the three primary colours used in normal colour film) but to green, red and near infrared. On processing, these colours are rendered as blue, green and red (hence 'false colour'), so that, for example, a green tank hiding in a green wood appears as a blue tank in a red wood.

4.3 Speed, contrast and resolution

The response of a photographic film to incident radiation is characterised most simply by these three parameters. The *speed* of a film refers to the length of time for which the film must be exposed to a given irradiance, in order to achieve a significant change in its opacity after processing. The speed is usually quantified by an ASA (American Standards Association) or DIN (Deutsche Industrie Norm) number, with large numbers corresponding to fast films and short exposure times.

The speed of a film describes its response to light of a single intensity. The *contrast*, on the other hand, describes the effect of changing the irradiance (or the exposure time). If a small change produces a large change in the opacity of the processed film, the film is said to have a high contrast, and conversely.

The response of a film to variations in exposure is summarised graphically by the *characteristic curve* (or Hurter–Driffield curve). This is a graph of the optical density D of the processed film against the logarithm of the exposure (X) to which it has been subjected. The optical density is defined by

$$D = -\log_{10}T$$

where T is the intensity transmission ratio of the film, so that

$$D = 0.434\tau$$

where τ is the opacity as defined in section 3.4.2. The exposure X is defined as the product of the irradiance E at the film and the exposure time Δt, so that its units are $Wm^{-2}s$ (or Jm^{-2}). There are, however, two important points to be made concerning exposure. The first is that its use implies that it is only the product $E\Delta t$ which is significant, e.g. that simultaneously doubling E and halving Δt will not change the appearance of the processed film. This assumption is known as *reciprocity*, and under extreme conditions (Δt very large or very small) it breaks down, which is a phenomenon known as *reciprocity failure*.

The second point to be made concerning exposure is that *photometric units* are often used in place of the radiometric units (Wm^{-2} for irradiance and exitance, and so on) with which we have dealt until now. Photometric units are weighted with respect to the nominal spectral sensitivity of the human eye (see appendix, and e.g. Kaye & Laby 1973), which is approximated by panchromatic film. The photometric quantity corresponding to irradiance is called *illuminance*, and is measured in *lux* (see appendix), so that the photometric unit of exposure is the lux-second.

A typical characteristic curve is shown in fig. 4.2. The horizontal (asymptotic) parts of the curve are labelled D_{fog} and D_{sat}. D_{fog} is the fog level, caused by a small proportion of the grains which have not been exposed being developed nevertheless. This may have a number of causes, including failure of the development process to distinguish between grains with and without development centres, and the generation of development centres by stray gamma-rays or X-rays. D_{sat} is the saturation level, corresponding to development of all the grains in the film. It is clear that if a film is exposed and processed so as to achieve either of these values of D, very little information is available about the exposure X.

The speed of the film is indicated by the value of X at which the curve of fig. 4.2 begins to rise. The slope γ of the useful, central, part of the curve is a measure of the contrast. A film with high contrast will have a large value of γ. Most films have $\gamma \approx 2$. Note that a large value of γ will reduce the input *dynamic range* of the film, that is, the range of values over which X may vary while still producing a useful change in the optical density D.

We may go some way towards understanding the shape of the characteristic curve by using a very simple model. If the grains present a cross-sectional area A to the incident radiation, and are suspended in the gelatin such that there are N grains per unit area of the film, we would expect (assuming that a developed grain is completely opaque) that

$$D_{sat} = -\log_{10}(1 - AN)$$

If the film is subjected to an irradiance E for a time Δt, the average number of photons which strike a particular grain is

$$\frac{AE\Delta t}{h\nu}$$

$$= \frac{AX}{h\nu}$$

where h is Planck's constant and ν is the frequency of the radiation. We thus expect the probability of a given grain containing a development centre, and hence the opacity of the developed film, to be some increasing function of AX. If the grain size A is reduced, a larger exposure X will be needed to achieve the same opacity and the film will be slower. If the grain size A is subject to a wide variation in the film, the input dynamic range will be increased and thus the film will have a low contrast. It is important

Fig. 4.2. The characteristic curve of a typical medium-speed film.

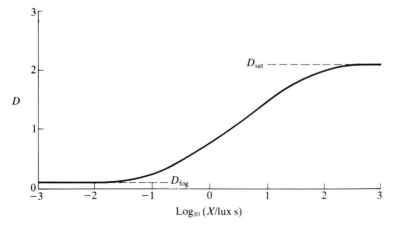

to note, however, that the processing applied to the exposed film can also have a pronounced effect on the shape of the characteristic curve, so that the latter cannot truly be said to be a property of the film alone.

   Resolution is, roughly speaking, the ability of a remote sensing system to distinguish an extended object from a point. It is one of the most important parameters describing the performance of a system, and we shall return to it in later chapters. For photographic systems, the resolution is normally expressed in line-pairs (lp) per unit length. That is to say, an object resembling fig. 4.3 is photographed, and said to be resolved if the bar pattern is recognizably reproduced on the negative. The resolution is then the greatest number of these bars, per unit length, which can be resolved *on the negative*. It is evident that if the geometry of the photographic system is known, this may be converted into a resolution, again in line-pairs per unit length, at the target. For example, if the imaging system consists of a lens of focal length f which is at a distance H from the target, and the resolution on the film plane is l line-pairs per unit length, it will be lf/H in the target plane. This is conventionally regarded as equivalent to the ability to resolve two *points* separated by a distance r_g, where

$$r_g = H/2lf \qquad\qquad (4.1)$$

Thus, if $l = 10^5 \, \mathrm{m}^{-1}$, $H = 2000 \, \mathrm{m}$ and $f = 150 \, \mathrm{mm}$, we find that $r_g = 6.7 \, \mathrm{cm}$.

   The resolution of an optical system is a combination of the performances of the lens and the film. The resolution of the film itself depends on the grain size, which, as we have said, is typically 1–$10 \, \mu\mathrm{m}$. A medium resolution film will be capable of recording $40 \, \mathrm{lp/mm}$, and a good reconnaissance film $100 \, \mathrm{lp/mm}$, but the latter will be correspondingly slow because of the fine grain size.

   We saw in chapter 2 that diffraction at an aperture of diameter D broadens plane parallel radiation into a cone of angle $\approx \lambda/D$ radians (2.32), and this also sets a limit on the resolution of the system. If

Fig. 4.3. Typical object used to determine the resolution of a photographic system in terms of line-pairs per unit length.

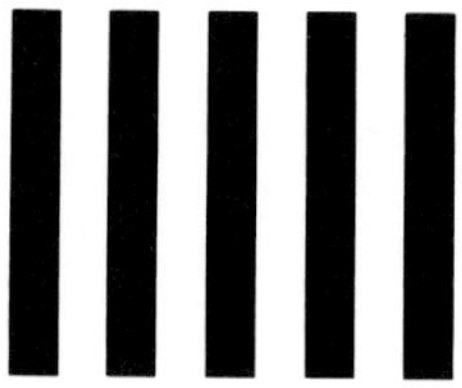

$\lambda = 0.5\,\mu\text{m}$ and $D = 5\,\text{cm}$, this angle is about 10^{-5} radians. This spreading out by the radiation will cause a point object to be imaged as a blurred spot. If the focal length of the system is 150 mm, the spot will have a diameter of about $1.5\,\mu\text{m}$. By comparing this with equation (4.1), we see that this is equivalent to a resolution of about 300 lp/mm. In this case, the resolution limit imposed by the lens is finer than that imposed by the film, so that the latter will dominate the system's performance.

It is conventional in photographic interpretation to distinguish between the ground resolution needed to *detect* a feature, to *identify* it, and to *analyse* it. As an example, the resolution needed to detect a road is about 10 m, whereas the resolution needed to identify it is about 2 m and that to analyse it about 0.2 m. Large geographical features such as shorelines, mountains and major rivers usually require a resolution for identification of about 15 m, and settled areas about 5 m.

4.3.1 The modulation transfer function

The use of line-pairs per unit length as a measure of film, or system, resolution, is a comparatively crude method of specifying the spatial fidelity of the representation. A more informative indication is given by the *modulation transfer function* (MTF). This describes the ability of the film to record sinusoidal variations in intensity, as a function of their spatial frequency. Because of the fact, mentioned but not proved in section 2.3, that any function can be constructed from a (possibly infinite) set of sinusoids of different amplitude and frequency, this approach contains in principle all information about the spatial response of the film.

Fig. 4.4 shows a sinusoidal variation of intensity with position. Apart

Fig. 4.4. A sinusoidal variation of intensity I with position x. This form of variation is characterised by its modulation $m = (I_{max} - I_{min})/(I_{max} + I_{min})$, and its spatial frequency q.

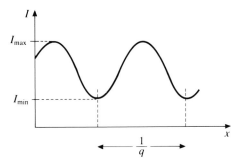

from its phase, which we shall ignore, this function is entirely characterised by its spatial frequency q (cycles per unit length -- note that this differs
from the usual definition of spatial frequency, which is measured in
radians per unit length and is therefore 2π times larger than this), and its
modulation m, defined as

$$m = (I_{max} - I_{min})/(I_{max} + I_{min}) \qquad\qquad (4.2)$$

The MTF of a photographic system is defined as the ratio of the output
modulation, i.e. that produced on the film, to the input modulation, i.e.
that of the target. It is a function of the spatial frequency q, and for
photographic systems this is conventionally defined as spatial frequency
in the film plane. Fig. 4.5 shows typical MTFs for coarse and fine-grained
films.

MTFs can be defined separately for the components of an optical
system (film, lens, etc.), and combined by multiplying them together. This
gives them a great advantage, compared with the line-pair criterion, in
assessing the performance of a complicated optical system.

4.4 Photographic optics

In this section we shall consider briefly the optics of photographic
systems. It is assumed that the reader is familiar with the simple theory of
image formation by single lenses; if not, he is referred to any elementary
textbook on optics, or to the book by Engel (1968) already mentioned.
Let us consider first a system with a single lens of focal length f (fig. 4.6).

Fig. 4.5. Modulation transfer functions of typical photographic films. (*a*) low
resolution, (*b*) high resolution.

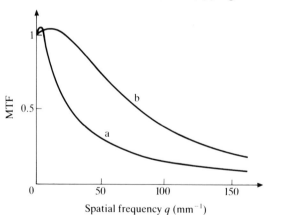

The object distance u and the image distance v are related by

$$1/u + 1/v = 1/f \qquad (4.3)$$

and it is clear that an object subtending an angle θ will produce an image of length $v\theta$ (assuming θ to be very much less than one radian). As the object distance u tends to infinity, v tends to the focal length f and the image size to $f\theta$. Clearly in all practical cases of aerial photography, this will be a justifiable approximation.

Let us consider the brightness of the image produced in this case. Consider an object of uniform exitance such that the incoming radiance at the lens is L. Let us suppose that the object subtends a solid angle Ω at the lens. The irradiance at the lens is thus ΩL, and the total flux intercepted by the lens is $(\pi D^2/4)\Omega L$, where D is the diameter of the lens. This flux will, assuming that there are no losses in the lens, be distributed over an area Ωf^2 of the image, giving an irradiance at the film plane of $E_{\text{film}} = (\pi D^2/4)L/f^2$. Thus

$$E_{\text{film}}/L = (\pi/4)(D/f)^2 \qquad (4.4)$$

The brightness of the image is thus determined by the ratio f/D, which is termed the *f/number* of the lens. (This rather curious designation shows that the number in question is the ratio of f to D.) Clearly, the smaller the f/number, the larger the lens and the brighter the image. Lenses can be constructed with f/numbers as small as about 1, although most lenses used in aerial mapping have f/numbers in the range 5 to 10.

4.4.1 Compound lenses

Most aerial photographic systems use compound lenses. These have the advantage of giving increased focal length without making the lens

Fig. 4.6. Formation of an image by a single lens.

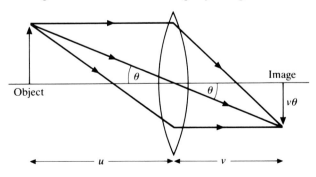

assembly any larger physically. Fig. 4.7 illustrates a compound lens using two converging lenses, with focal lengths f_1 and f_2 respectively. Again we shall assume that the object is located at infinity, so that the first lens (the *objective lens*) forms an image of it in its focal plane. From fig. 4.7, it is clear that the combined effect of the two lenses is equivalent to that of a single lens of focal length $f_1 v_2/u_2$, that is, the focal length of the objective multiplied by the magnification of the second lens. All of the remarks we have made about single-lens systems are thus still valid, so long as we substitute this effective focal length for the simple focal length f.

In fact, an even greater saving of space is made if the second lens of the compound system is of the diverging, rather than the converging, type. This is the usual construction of telephoto lenses.

4.4.2 Scale and coverage of the image

The scale of a map, and, by extension, of an aerial photograph, is the number less than unity which expresses the ratio of the size of the representation of an object on the map to the size of the real object. 'Large scale' is generally taken to mean greater than $1/50\,000$ and 'small scale' less than $1/500\,000$. It is clear, from simple geometrical considerations, that the scale of the negative of an aerial photograph is

$$f/H$$

where f is the focal length and H the distance from the lens to the ground. Naturally, prints may be produced by enlargement or reduction from the negative. What is fixed, however, is the geographical coverage of the

Fig. 4.7. Formation of an image (of an object at infinity) by a compound lens.

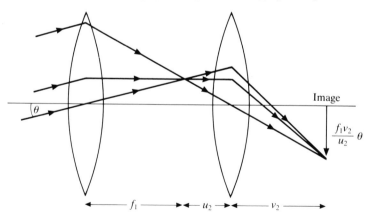

image. Inspection of fig. 4.8 shows that if the negative has a size s, the corresponding ground coverage will be

$$D = sH/f \qquad (4.5)$$

The size s ranges typically from 35 mm to about 30 cm, used in 'large format' cameras carried by, for example, the space shuttle. Since the ratio s/f is of the order of unity in mapping applications, the coverage of an image will be roughly equal to the height from which it was obtained.

Note that all of our considerations of ground scale, ground resolution and ground coverage have been derived for the case of *vertical* aerial photography. *Oblique* photography (figs. 4.9 and 4.10) gives much greater coverage, but at the expense of variable resolution and scale. For this reason it is generally unsuitable for mapping or for quantitative stereoscopic applications.

4.5 Stereophotography

It is at first sight surprising that aerial photographs contain information about the heights of surface features. In order to extract the maximum of information, a pair of photographs must be used, but a single photograph contains some information in the form of *relief displacement*.

Consider a vertical aerial photograph of a vertical object of height h, as shown in fig. 4.11. The object is located at a distance a from the camera's nadir point, and the camera (whose focal length is f) is located at a height H above the ground. It is clear from fig. 4.11 that, as long as $a \neq 0$ (that is,

Fig. 4.8. Geometric construction to find the coverage D of a vertical aerial photograph.

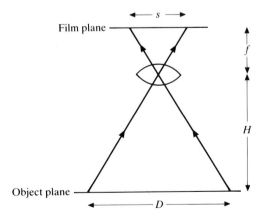

Fig. 4.9. The oblique aerial photograph obtains greater coverage at the expense of variable scale and resolution.

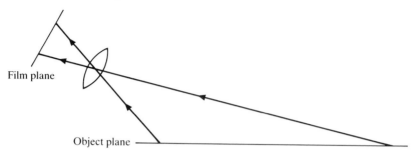

Fig. 4.10. An oblique aerial photograph, showing wide coverage and variable scale. The image is of Cambridge, England. (Cambridge University Collection: Copyright reserved).

the object is not directly below the camera), the head and foot of the object are imaged at different points on the film. The distance h', which is the projection of the height h onto the film, is called the relief displacement. Simple geometry shows it to be

$$h' = hfa/H(H-h) = ha'/(H-h) \qquad (4.6)$$

which is approximately ha'/H if $H \gg h$. Thus if we have a vertical object (e.g. a tower, or the side of a building), we can determine its height from a single aerial photograph, provided we know the height from which the photograph was obtained.

If we assume that the relief displacement h' can be measured with an accuracy of $\Delta h'$, a combination of (4.5) and (4.6) shows that the corresponding accuracy Δh to which the height can be determined is given by

$$\Delta h = 2H\Delta h'/s \qquad (4.7)$$

where s is again the film size. If we take $s = 230$ mm (i.e. 9 inches, a common size for aerial photography) and assume that $\Delta h' = 0.1$ mm, we find that Δh is close to $H/1000$. This is a common and useful rule of thumb.

The method of relief displacement, relying on the measurement of h', depends on the image revealing both the top and the bottom (i.e. the point vertically below the top) of the object. While this is possible for some buildings (and only if the photograph is taken from a suitable location), it will clearly not work in the case of topographic features. In such cases, we need to use a *pair* of photographs taken from different locations.

Fig. 4.11. Relief displacement. The top and bottom of a vertical object are imaged at different places on the film.

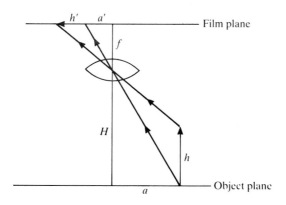

Fig. 4.12 shows the relevant geometry of the situation. It illustrates real rays from the mountain peak, and 'imaginary' rays from the mountain base, vertically below the peak. Using (4.6), we find that

$$h_1' = ha_1'/(H-h)$$
$$h_2' = ha_2'/(H-h)$$

and

$$a_1' = a_1 f/H$$
$$a_2' = a_2 f/H$$

Although h' and a' are not measurable separately, their sum, corresponding to the position of the image of the mountain peak, is measurable. In particular, we shall consider the *difference* in this quantity between the two photographs, by putting $\Delta x = (h_1' + a_1') - (h_2' + a_2')$. This is given by

$$\Delta x = (a_1' - a_2')H/(H-h)$$

i.e.

$$\Delta x = bf/(H-h) \tag{4.8}$$

where b is the *baseline*, or distance between the two points from which the photographs were taken. Thus, without being able to see the base of the

Fig. 4.12. Stereophotography. The height h of the mountain can be determined by analysis of two photographs, even though its base is not visible in either.

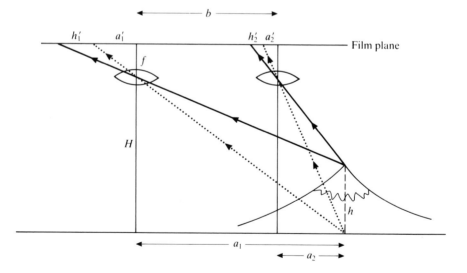

mountain in either photograph, we can deduce its height so long as it is visible on both photographs. The accuracy achievable by this technique is the same as that for relief displacement, since the method is fundamentally the same. Fig. 4.13 shows a pair of stereophotographs.

The actual calculation of heights from a pair of stereophotographs may be made by measuring from prints, or by digitising them and analysing the results in a computer. However, a common way of analysing stereophotographs where quantitative height information is not required is to view them through an instrument which presents one image to each eye. The human brain is familiar with the interpretation of slightly different perspectives seen by the two eyes; indeed this is the normal

Fig. 4.13. A pair of stereophotographs. Each shows the phenomenon of relief displacement (e.g. the cooling towers left centre), and by viewing the pair together an impression of three-dimensional relief can be obtained. The images were recorded from a height of approximately 900 m, and have a coverage of about 1.3×0.9 km. They show a motorway interchange in Birmingham, England. (Cambridge University Collection: Copyright reserved)

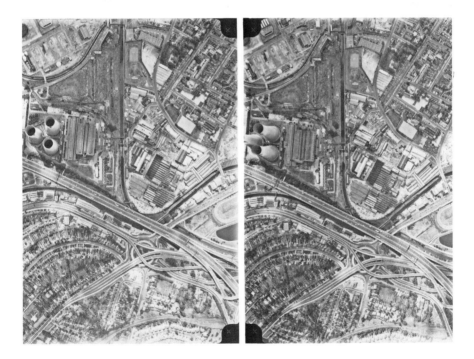

functioning of binocular vision. The object of the stereoscopic viewer, then, is to fool the brain into reconstructing a sensation of three-dimensionality. The most sophisticated viewers are genuinely quantitative in that it is possible to determine the heights of objects from them.

An alternative, related method of viewing stereophotographic pairs is by means of an *anaglyph*. Here, the two images are portrayed in different colours, one red and one green, and viewed through coloured filters arranged so that each eye sees only one image. If this is done by projecting the two images optically, the method should work as well as a conventional stereo viewer. However, what is often done for cheapness is to *print* the two images on a single sheet of paper using different coloured inks, and the inability of the printing process to represent correctly the addition of colours means that the resulting three-dimensional effect is only approximate. Plate 3 shows a typical anaglyph, although it has in fact been derived from an electro-optical sensor (see chapter 5) rather than by photography.

Most people's eyes are about 7 or 8 cm apart, and focus comfortably on objects held about 50 cm distant. Because of this, stereo photographs recorded with the same ratio (1/6) of baseline to height will appear to have the correct perspective when interpreted by the brain. Conversely, if the ratio b/H is larger than 1/6, the reconstructed image will appear to suffer from a proportional vertical exaggeration. Note that this is merely an artefact of the way the brain interprets parallax, and is not quantitatively inherent in the data. However, such exaggeration is often useful, for example in identifying and interpreting geomorphological features.

4.6 Atmospheric effects

We have discussed in section 3.4 the effects of the atmosphere on electromagnetic signals propagating through it. In section 3.4.3 we saw that atmospheric turbulence will impose a limit on the angular resolution of a system, although this is unlikely to be important except in the case of high-magnification systems of high intrinsic resolution. A more important atmospheric effect is the reduction in image contrast due to atmospheric scattering.

The contrast C of a scene, or of some part of it, is defined as

$$C = (L_{max} - L_{min})/(L_{max} + L_{min}) \tag{4.9}$$

(although note that other definitions, such as L_{max}/L_{min}, are also used), where L_{max} and L_{min} are respectively the maximum and minimum luminances (or, in radiometric terms, radiances) of the scene. Absorption *by* the atmosphere will reduce the apparent luminances, whereas the corresponding reradiation *from* the atmosphere itself will increase them.

Let us write E_S for the illuminance at the earth's surface due to skylight (and direct sunlight, if present), E_A for the (upwards) illuminance of the atmosphere itself, T for the intensity transmittance of the atmosphere and r for the reflectance of the scene. In this case, the contrast of the scene is

$$(r_{max} - r_{min})/(r_{max} + r_{min})$$

The detected illuminance due to the scene will be $E_S r T$, so that the observed contrast will be

$$(r_{max} - r_{min})/(r_{max} + r_{min} + 2E_A/E_S T)$$

The scene contrast is thus reduced, and may possibly be reduced below the level which can be detected on the film.

Fig. 4.14 shows the typical sunlight, skylight and atmospheric il-luminances, with the blue end of the spectrum filtered out, for an observation made above approximately one atmospheric scale height (i.e. about 8 km), as functions of the solar elevation (see also the appendix).

The atmospheric transmittance T may typically be taken as 0.8. As an example, let us consider the contrast of two objects with reflectances of 0.1 and 0.2. The intrinsic scene contrast is clearly 0.33, from (4.9). At a solar elevation of 50° we have $E_A \approx 2.0 \times 10^3$ lux and $E_S \approx 77 \times 10^3$ lux (if the sun is shining), so that the observed contrast will be reduced to 0.27. When the solar elevation is 10°, $E_A \approx 1.7 \times 10^3$ lux and $E_S \approx 12 \times 10^3$ lux, giving a contrast of only 0.15.

This effect is stronger at wavelengths where scattering or absorption are strong, and conversely. Thus, the scene contrast will generally be lower at the blue end of the spectrum than at the red or infrared end.

4.7 Applications of aerial photography

The applications of aerial photography are in general well known, and (with the exception of the use of infrared film) the correspondence between a photographic image and the perception provided by the human eye and brain is sufficiently great that most applications have in any case an intuitive feel about them. The main advantages of photography as a remote sensing technique are that it is controllable and inexpensive, and that photographic optical systems can be made with sufficient precision that quantitative spatial information may be obtained from the resulting images. In this way, photography has found widespread application in mapping and surveying, for example in geology and hydrology, in terrain analysis, field mapping, regional planning, the study of crop types and diseases and so on. Colour film, though expensive, is also widely used, especially in cases where vegetation is studied (for example agriculture, forestry and ecology) but also in geomorphology, hydrology and oceanography.

Black and white near infrared film has proved especially useful in studying soil moisture and erosion, and in archaeological surveying. FCIR film has found application in the classification of urban areas, in monitoring soil moisture and performing animal censuses, in disaster assessment, and (particularly) in vegetation mapping.

Fig. 4.14. Clear-sky illuminances as functions of solar altitude. (*a*) Illuminance at the earth's surface due to direct sunlight. (*b*) Illuminance at the earth's surface due to skylight. (*c*) Illuminance at top of atmosphere due to upwelling skylight.

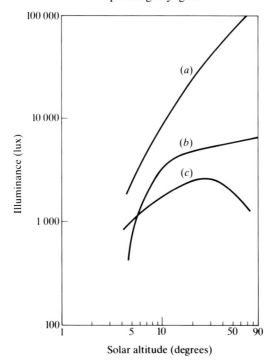

Problems

1. Prove (4.6).
2. A vertical aerial photograph reveals a tall building. The foot of one corner of the building has (x,y) coordinates (42.7, 88.2) (both measured in mm from the lower left-hand corner of the negative), and the top of the same edge appears at (26.6, 82.6). The corresponding coordinates for an adjacent edge of the building are (42.7, 103.6) and (26.6, 100.8). Given that the camera's focal length was 88 mm and the aircraft's altitude 212 m, find (*a*) the coordinates of the photograph's principal point (directly below the camera), (*b*) the height, and (*c*) the width, of the building.
3. A pair of vertical stereophotographs is examined. It is found that if the negatives are superimposed so that points at sea level coincide, the images of a mountain peak are displaced by 4.3 mm from one another. Given that the aircraft's altitude was 5000 m above sea level, the camera's focal length was 100 mm and the baseline was 1 km, find the height of the mountain.

5

Electro-optical systems

5.1 Introduction

In chapter 4 we discussed photographic systems, which stand somewhat apart from the types of system which will be considered in chapters 5 to 8. In the case of photographic systems the radiation is detected through a chemical process, whereas in the systems which we shall now consider it is converted into an electrical signal which can be detected, amplified and subsequently further processed electronically. This clearly has many advantages, not least of which is the comparative ease with which the detected data may be transmitted as a modulated radio signal and recorded on magnetic tape.

In this chapter we shall consider such electro-optical systems used for detecting visible and infrared radiation. It is convenient to group visible, near infrared and thermal infrared systems together since, as we shall see, much of the detector technology is the same in these wavebands. We shall also discuss in this chapter some of the more important uses to which these systems are put.

5.2 Scanning and non-scanning systems

We may regard aerial photography as a non-scanning imaging system. It is clearly an imaging system in the sense defined in chapter 1, in that it forms a two-dimensional representation of the two-dimensional brightness distribution over the target. It is non-scanning in that the entire image is recorded virtually instantaneously on a two-dimensional array (the negative). This is achieved using a more or less analogue method, recording a continuous variation of intensity over all points on the negative. In fact, of course, individual grains are either converted to silver or they are not, so that at the microscopic level the photograph has more in common with digital systems. The statistical uncertainties inherent in the process mean that, at the macroscopic level, it may be regarded as an analogue process. However, we could easily imagine a system in which the negative was replaced by a two-dimensional array of very many detectors, each of which would generate an electrical signal dependent on the intensity of the light falling upon it. These signals could then be recorded and subsequently 'played back' to view the image (see fig. 5.1).

This device would evidently be functionally equivalent to a photograph, although with the desirable advantage that data could be transmitted electronically and processed in a computer. Such devices do in fact exist, in the form of *two-dimensional charge-coupled devices* (CCDs). A CCD consists in general of a one- or two-dimensional array of sensitive elements. Exposure to electromagnetic radiation generates an electric charge, proportional to the incident energy, at each element. The great advantage of the CCD is that these charges can be passed from one element to the next, under suitable electronic control, and 'read out' at the edge of the array (i.e. it is a form of shift register). In this way, the number

Fig. 5.1. Schematic view of a non-scanning electro-optical imaging system. The lens focusses incoming radiation onto the array of detectors. The detectors generate electronic signals which are dependent on the radiation intensity, and these signals are processed and stored. Such a system would be functionally equivalent to a photographic system of camera and film.

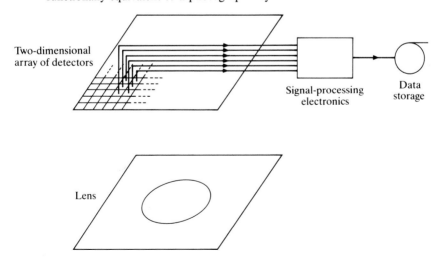

Two-dimensional array of detectors

Signal-processing electronics

Data storage

Lens

of electrical connexions needed for a rectangular array of n by m elements is $(n+m)$ rather than $(n \times m)$. Arrays of typically 1000×1000 elements, each of the order of $10 \ \mu m$ in size, have been constructed, and it seems likely that the use of CCDs in spaceborne and airborne imaging of the earth's surface will increase significantly.

The complexity and cost of a system with a million discrete detecting elements has however caused a simpler solution to be sought until now. It is simpler to use a one-dimensional array of detectors, or only a single detector, and to steer the direction from which this detector (or array of detectors) gathers its information, in order to build up the required two-dimensional array of data. This is the basis of scanning systems.

5.3 Operation of a typical scanning system

The commonest type of scanner in operation is probably the 'whiskbroom', line-scanner or single-element scanner. This has only one detector, or rather, one sensor for each spectral band to which the instrument responds. The area on the surface from which radiation is gathered by this detector is scanned in one dimension, usually by means of a rotating mirror. Scanning in the perpendicular direction is achieved by using the forward motion of the platform (e.g. aircraft or satellite) carrying the instrument.

Fig. 5.2 shows, in a simplified form, the construction of a typical whiskbroom system. It is in fact a diagram of the LANDSAT multispectral scanner (MSS) which, as its name suggests, has a number of spectral bands of sensitivity in the visible and infrared. The beam is scanned sideways by a rotating mirror, and focussed into a parallel beam by mirrors. (Note that although mirrors are often used in scanning systems, the optical concepts discussed in chapter 4 may still be applied.) This beam is then separated into visible and near infrared components on the one hand, and thermal infrared on the other, and these components are detected using appropriate optoelectronic devices. The electronic signals derived from them are finally formatted and recorded or transmitted.

It will be apparent that care must be taken to match the speed of side-to-side scanning to that of the forward motion. If the former is too small or the latter too large, strips of the surface will not be detected. The ideal relationship between the two is governed by the width Δx of the area of sensitivity (see fig. 5.2), which is usually called the *instantaneous field of view* (IFOV). (Note that, for all scanners, we should distinguish carefully between the IFOV and the FOV, or field of view. The scanning process essentially synthesises the FOV from many different positionings of the IFOV.) Clearly, if the platform is not to advance by a distance of more than Δx in the time ΔT for one line to be scanned, the forward speed v must satisfy

$$v < \Delta x / \Delta T \qquad (5.1)$$

The LANDSAT MSS system, for example, achieves an IFOV $\Delta x = 79$ m, and has an equivalent ground speed $v = 6.46$ km s^{-1}. The scan time ΔT must therefore be at most 12.2 ms. In fact, the MSS scans 6 lines simultaneously in a time of 73 ms, so that the effective value of ΔT for each line is matched at 12.2 ms.

We should also note at this point that the lines scanned across the surface will not be perpendicular to the direction in which the mirror is scanned. During the time ΔT in which the line is scanned, the platform advances by a distance $v\Delta T$. If the width of the scanned line is w, we can see by simple trigonometry that the line will be oriented at an angle θ to the direction in which the mirror is scanned, where

$$\tan \theta = v\Delta T / w \qquad (5.2)$$

Fig. 5.2. Schematic view of a whiskbroom sensor. The IFOV (shaded area) is scanned from side to side by the oscillating mirror. Scanning is achieved in the perpendicular direction by the motion of the scanner as it is carried by the aircraft or satellite.

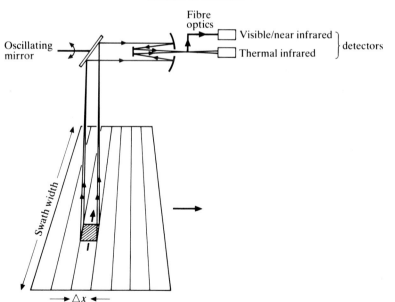

This clearly represents a form of geometric distortion in the image. The way in which such distortion is removed is discussed in chapter 10.

We have introduced the concept of the IFOV, and clearly we ought to discuss the factors that determine it. We have already seen in chapter 2 that the *angular* resolution has a lower limit, set by diffraction, of about λ/D, where λ is the wavelength and D is the diameter of the first obstruction encountered by the radiation. This will usually be the objective lens or mirror of the system. Applying this simple formula, we can write

$$\text{IFOV}_{\text{min}} \approx \lambda H/D \qquad\qquad (5.3)$$

where H is the height of the platform above the surface. An exact version of (5.3) can be derived by considering in greater detail the diffraction of the radiation. It is important to note that this equation gives the *minimum* size of the IFOV. Since the detectors themselves have a finite size, and since the signal derived from a detector is, in effect, an average of the radiation intensity over the entire detector, the IFOV cannot be smaller than the size of the detector *projected onto the surface*. This is the same point which we made about the resolution of a photographic system in chapter 4. For a simple optical system characterised by a focal length f, the angular IFOV will be a/f where a is the detector size, so

$$\text{IFOV}_{\text{min}} = Ha/f \qquad\qquad (5.4)$$

Thus, in practice (and approximately speaking), the IFOV will be whichever is the larger of (5.3) and (5.4). However, we note that it will always be proportional to the height H of the platform. Combining this observation with (5.1), we can see that the height and speed of the platform, and the scan rate ΔT, cannot be varied independently. This imposes operational considerations for airborne applications, and design considerations for satellite-borne applications (in which the relationship between v and H is fixed), to which we shall refer again in chapter 9.

The foregoing discussion has outlined the origin of the IFOV in remote sensing systems. It must be noted that the IFOV and the spatial resolution are not necessarily identical, although in an optimally designed system they are likely to be similar in size. The question of what, precisely, is meant by spatial resolution in the case of a scanning system is in fact rather difficult to answer. A thorough review of the various possible interpretations of the term has been presented by Forshaw *et al.* (1983).

We have discussed at some length the operation of a line-scanner or 'whiskbroom' scanner, in which the IFOV can be regarded as, in effect, zero-dimensional. Before turning to a consideration of the design of detectors, we shall briefly mention the 'pushbroom' scanner, in which the IFOV is one-dimensional. In this type of instrument there is a linear array of sensors, thus achieving quasi-instantaneous 'scanning' in the across-track direction. As before, scanning in the perpendicular direction is achieved by utilising the forward motion of the platform. Such an instrument, therefore, has no moving parts, and is in consequence potentially smaller, lighter, cheaper and more reliable than the line-scanner. The disadvantage is that there are of the order of a thousand detectors per spectral band, instead of just a few (6 in the case of the LANDSAT MSS), each of which requires calibration. This type of system is thus in one sense more complicated than the line-scanner, and in another sense simpler. It has been adopted for the HRV (high resolution visible) sensor flown on board the French SPOT-1 satellite.

It is clear that two-dimensional (CCD) arrays, one-dimensional arrays (pushbroom systems) and 'zero-dimensional arrays' (whiskbroom systems) have differing strengths and weaknesses. A two-dimensional system allows a given surface resolution element to remain in view for the longest possible time, thus increasing the radiometric sensitivity. It is also comparatively insensitive to irregularities in the motion of the platform (see chapter 9), giving a good geometric fidelity in the image, and, having no moving parts, it is mechanically robust. The disadvantages of such a system are its need for optics giving a wide field of view, the difficulty of providing adequate cooling for such an array if it is to be used for detecting thermal infrared radiation, the difficulty of observing more than one spectral channel, and the problem of calibrating large numbers of sensors.

Finally, we should note that a different scanning approach must be adopted for an instrument operated from a satellite in geostationary orbit (see chapter 9). In such a case, the instrument remains above a fixed point on the earth's equator. The usual method of scanning is a modified form of whiskbroom scanning, in which side-to-side (i.e. east–west) scanning is achieved by making the satellite rotate about its axis (which is aligned parallel to the earth's polar axis), and fore-and-aft (north–south) scanning is achieved by slowly rotating a mirror within the instrument. This is the technique adopted for the VISSR (visible and infrared spin-scan radiometer) carried by the METEOSAT and GOES meteorological satellites. The VISSR scans the earth's disc in 1821 strips at 100 rpm, with an IFOV at nadir of 2.4 km in the visible band and 5 km in the thermal infrared.

5.4 Detectors for visible and near-infrared radiation

We have already introduced the LANDSAT MSS system, in order to describe the operation of a line-scanner. The MSS records in four spectral bands (LANDSAT 3 also recorded a thermal infrared band), namely 0.5–0.6 μm, 0.6–0.7 μm, 0.7–0.8 μm and 0.8–1.1 μm. The *thematic mapper* (TM), a more accurate instrument carried by later LANDSATs, has in addition bands at 0.45–0.52 μm, 1.55–1.75 μm and 2.08–2.35 μm, as well as a thermal infrared band. In this section we shall discuss how sensitivity over these ranges of wavelength is achieved, although a discussion of the numerical values of the sensitivities is beyond our scope.

5.4.1 Photomultipliers

The visible bands normally use *photomultipliers* as detectors. The photomultiplier is a simple device, shown schematically in fig. 5.3.

An incident photon ejects an electron from the photocathode, by the

Fig. 5.3. A photomultiplier tube.

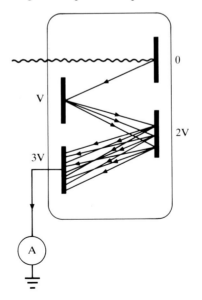

photoelectric effect. This electron is accelerated towards an intermediate electrode, at a more positive potential than the cathode, and the increased kinetic energy causes it to eject more than one electron from the intermediate electrode. This process is repeated several times, the number of electrons increasing at each step until a measurably large current is generated. The size of this current depends on the intensity of the incident radiation.

The minimum photon energy which can be detected by such a device is the *work function* of the material from which the photocathode is made. The work function W is defined as the energy difference between an electron *in vacuo* and an electron within the material. For metals, values of W lie typically in the range 2–5 eV, so that the maximum wavelength to which such devices are sensitive is about $0.6\,\mu m$. Mixtures of alkali metals, however, can have significantly smaller work functions, and sensitivity up to about 1 μm can be obtained. To define a narrow spectral band of sensitivity, the input radiation is filtered.

The photomultiplier is a very sensitive device with a rapid response time (of the order of 1 ns). Its main disadvantages are its mechanical fragility, and the high operating potential of about 1 kV.

5.4.2 Photodiodes

Radiation in the near-infrared region is usually detected using a photodiode. This is a semiconductor junction device, usually indium antimonide (InSb) or lead sulphide (PbS), in which an incident photon generates a current or voltage across the junction. The signal is proportional to the light intensity.

A semiconductor diode consists of two pieces of semiconducting crystal in intimate contact. One piece has been *doped* with a trace of impurity which gives rise to an excess of electrons, the other with an impurity giving a deficit of electrons. These are referred to as n-type and p-type material respectively, since the effective charge carriers are, respectively, negatively (electrons) and positively ('holes' – i.e. absences of electrons) charged. At the junction, holes from the p-type material diffuse into the n-type material, where they combine with the free electrons. A corresponding effect occurs also in the opposite direction, giving rise to a *depletion region* of very low conductivity, about 1 μm in width. Because there is now an excess of positive charge in the n-type material, and of negative charge in the p-type material, there is an electric field from n-type to p-type across the depletion region. This field inhibits the further diffusion of charges.

If an external electric field is applied from p-type to n-type (forward bias), the internal field is overcome to some extent and the depletion region is made narrower. A current flows, and its magnitude increases roughly exponentially with the applied voltage. If however the external field is applied in the opposite sense (reverse bias), the depletion region is made wider, and a much smaller current flows. Fig. 5.4 summarises the behaviour of a semiconductor diode.

If now the diode (with no external bias) is subjected to incident radiation, a photon may be able to create an electron–hole pair in the p-type material. If the electron diffuses into the depletion region it will be accelerated by the internal field into the n-type material, and the work done will appear as a voltage, proportional to the intensity of the radiation, across the diode. This is called the *photovoltaic mode* of operation. It has the advantage that no signal is generated when no radiation is incident upon the diode, and the disadvantage of a comparatively long response time of about 1 μs.

If on the other hand the diode is reverse biassed, increasing the width of the depletion region, the external field maintains the field within the depletion region, and the charge-carriers will generate a current. This is called the *photoconductive mode*, and it usually gives a faster response time than the photovoltaic mode (typically 1 ns). The reason for this is that the response time is determined by the capacitance of the depletion region, which is inversely proportional to its width.

Fig. 5.4. Behaviour of a semiconductor diode. Note the comparatively large current which is passed when the diode is forward biassed ($V > 0$), and the much smaller current when the diode is reverse biassed.

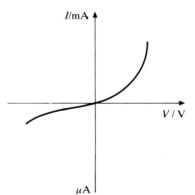

The maximum photon wavelength which can be detected by a photodiode is determined by the energy required to create an electron–hole pair. This is termed the *band-gap* of the semiconductor. Semiconductors such as germanium have relatively large band-gaps, giving sensitivity up to only 1.7 μm, but PbS responds up to about 3 μm and InSb to about 5 μm.

5.4.3 Vidicons

The vidicon is a type of television camera, and was the earliest electro-optical device to be used for obtaining visible-wavelength imagery from space. Fig. 5.5 shows schematically the design of a simple vidicon, consisting of a lens and shutter, and a sheet of photoconductive material coated on the side nearer the lens with a transparent conductor. The other side of the photoconductive sheet is illuminated by an electron beam which can be deflected electrostatically (in much the same way as in a cathode-ray oscilloscope) to impinge upon any part of the sheet. The mode of operation is as follows. First, with the shutter closed, the electron beam is used to coat the back of the photoconductive sheet with electrons. The electron beam is then turned off, and the shutter is opened and closed, leaving a distribution of charge on the sheet corresponding to the distribution of light intensity. Finally, the back of the sheet is scanned with the electron beam, resulting in an electric current (the read-out) from those parts which were illuminated.

Fig. 5.5. Design of a vidicon (schematic).

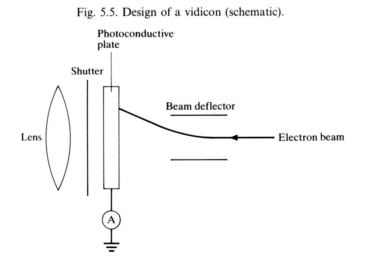

The early LANDSAT missions in fact used a device known as a return-beam vidicon (RBV), whose detailed operation is somewhat different from this, although the principles are similar.

5.5 Applications of visible and near-infrared imagery

Visible and near-infrared imagery, usually with quantitative or even calibrated brightness values, has found enormous application in remote sensing. Part of its popularity may be ascribed to the greater ease with which such data may be interpreted, since the wavelength range corresponds largely with the eye's range of sensitivity, but to a great extent also many important processes modulate the brightness in this range of wavelengths. In this section we cannot discuss these applications at great length, but it is appropriate at least to mention some of the more important uses to which such data are put. Plate 4 shows a typical satellite image acquired by an electro-optical scanner.

The development of biomass (total organic matter per unit area of the earth's surface) can be monitored, especially in conjunction with thermal infrared data. The ability to observe changes in the extent and state of health of vegetation is valuable, both for the commercial estimation of crop yields, and for the early prediction of drought and crop failure. In a somewhat similar manner, different types of tree can be distinguished in forestry. Regions of snow and ice can also be monitored, and the amount of runoff during the melting season estimated. In order to distinguish between cloud and snow, use of a channel at 1.5–1.8 μm is especially useful, since cloud has a high albedo in this region, whereas snow has a much smaller albedo (see Crane & Anderson 1984).

The depth of shallow bodies of water can be estimated by comparing the intensities of the radiation reaching the detector in each of two (or more) spectral bands. This method relies on the variation with wavelength of the attenuation coefficient, although because of uncertainties in this coefficient the method generally requires calibration. The best spectral bands to use are green and near-infrared, since these give the largest difference in attenuation.

Marine plankton concentrations (which probably delineate good fishing areas) are estimated by the increased reflectance, caused by the presence of chlorophyll, in the 0.4–0.5 μm band (see chapter 3). The effect is normally a small one, so that particularly sensitive instruments must be employed, and careful correction must be made for atmospheric effects. Over the open sea, suspended sediments can be ignored (so-called 'type I waters') in estimating chlorophyll concentrations, and the calculation is usually accurate to within a factor of two, but in shallow waters suspended sediments, especially those with significant reflexion in the yellow band, severely complicate the problem. Ocean currents have been estimated from spatial variations in turbidity. Gordon & Morel (1981) provide a good introductory discussion of these 'water colour' measurements, and Robinson (1985) treats the subject in great detail.

Clouds can be delineated and monitored, measuring the degree of cloud cover, its height and type, as an aid to weather forecasting. This is particularly valuable in the case of sensors with wide swaths, such as those carried by the polar orbiting meteorological satellites and, especially, the geostationary satellites. By tracking the motion of clouds, wind speeds can be obtained (Leese, Novak & Clarke 1971, Warren & Turner 1988).

In summary, the main uses of the spectral bands (defined purely for convenience) in the range $0.4–1.8\,\mu m$ are as follows:

$0.4–0.5\,\mu m$ mapping chlorophyll distributions and coastal waters; discriminating between coniferous and deciduous trees; determining soil type

$0.5–0.6\,\mu m$ bathymetry; monitoring marine sediments and healthy vegetation

$0.6–0.7\,\mu m$ discriminating between plant species; mapping geological and cultural features

$0.7–0.8\,\mu m$ performing biomass surveys; mapping and monitoring vegetation; discrimination of land/water boundaries

$0.8–1.1\,\mu m$ vegetation mapping

$1.6–1.8\,\mu m$ measuring vegetation moisture content; discrimination of snow and cloud

In the future, it is likely that narrower spectral bands will be defined (probably by using a dispersive element such as a prism and mechanically scanning across its output), so that reflectance spectra can be more truly characterised. This will naturally greatly increase the usefulness of the data, especially in geological applications where characteristic features often have rather narrow bandwidths (of the order of 30 nm – see chapter 3, and Goetz *et al.* 1985).

5.6 Detectors for thermal infrared radiation

Even more exotic semiconductors than InSb can be found, with smaller values of band-gap and consequently higher values of maximum wavelength. In particular, mercury cadmium telluride ($Hg_{0.2}Cd_{0.8}Te$)

can be operated, usually in the photoconductive mode, up to about 15 μm, which adequately covers the thermal infrared region of the electromagnetic spectrum. It is of course desirable to cool any detector of thermal infrared, usually using liquid nitrogen (at 77 K), so that excessive thermal infrared radiation is not generated by the sensor itself. We should note also that glass is opaque to wavelengths longer than about 2 μm, so that other materials must be used for the construction of lenses and other optical elements.

As well as the photodiode, there are three other important types of thermal infrared detector, which we shall briefly describe. Further details may be found in, for example, Anderson & Wilson (1984).

5.6.1 Bolometers

The bolometer is a simple device which consists, in essence, of a material whose resistance varies with temperature. The effect of incident radiation in the thermal infrared region is to cause a rise in temperature, which can then be detected. The detecting elements may be platinum strips (whose resistance increases with temperature), or *thermistors* (semiconducting devices, usually constructed from mixed metal oxides. The resistance decreases with temperature as a result of the increased density of charge carriers). The bolometer's main disadvantage is its long response time, typically 10 ms.

5.6.2 Thermopiles

A thermopile is a series of thermocouples. Each of these uses the *Seebeck effect*, in which a potential difference is generated across a pair of junctions between dissimilar metals when the junctions are held at different temperatures. Like the bolometer, the thermopile has a long response time (of the order of 10 ms), and in addition it is not particularly sensitive. It is mechanically somewhat delicate, but responds well to wavelengths up to about 30 μm.

5.6.3 Pyroelectric devices

A pyroelectric detector is, in essence, a crystal which undergoes a redistribution of internal charge distribution as a result of a change in its temperature. Charge separation occurs at the surfaces of the crystal, resulting in a potential difference which can be amplified and detected.

This is usually done by 'chopping' the incident radiation (that is, interrupting it periodically) at about 1 kHz, and measuring the component of the output signal which alternates at the chopping frequency. The great advantage of pyroelectric detectors is that they can respond very quickly (down to 1 ns in some cases) to temperature changes, and are sensitive to wavelengths up to about 70 μm.

5.7 Applications of thermal infrared remote sensing

Two principal wavebands are used for thermal infrared remote sensing, separated by the atmospheric water vapour absorption bands at 5–6 μm. These are the 3–5 μm band and the 8–14 μm band. The intrinsic sensitivity of a radiometric observation at a given wavelength and temperature can be defined in terms of the Planck radiation function (2.44) as

$$S = (1/L_\lambda)\partial L_\lambda/\partial T \tag{5.5}$$

since this is the fractional change in the detected power for a unit change in the black-body temperature of the source. The 3–5 μm band is the more sensitive to changes in temperature at about 300 K, since the value of S is greater at 4 μm than at 10 μm, but the 8–14 μm band is more often used because the contribution from reflected sunlight is very much less (see Stewart 1985 for a discussion of this point). The 3–5 μm band has also proved useful, though, especially in the study of volcanoes and for night-time oceanography. The 8–14 μm band contains the peak of the black-body radiation spectrum for temperatures between about 210 and 360 K (-70 to $+90\,°C$).

Broadly speaking, we may classify applications of thermal infrared remote sensing into those in which the surface temperature is governed by man-made sources of heat, and those in which it is governed by solar radiation. In the former case, the technique has been used from airborne platforms for determining heat losses from buildings and other engineering structures. It is most useful to perform this kind of observation just before dawn, so that the effects of differential solar heating will have had the greatest possible time to decay.

In the latter case, thermal infrared remote sensing has been used for identifying soil moisture and measuring water stress, for identifying crop types and frost hollows, and so on. Two particularly important applications will be described in greater detail: measuring the surface temperature of the sea, and identifying geological materials by their thermal inertia.

5.7.1 Sea surface temperature

The sea surface temperature (SST) is a quantity of obvious oceanographic and meteorological importance. It can be deduced in a fairly straightforward way from calibrated thermal infrared data, since the emissivity of a pure water surface is a well defined quantity with a value of 0.993 (see fig. 5.6). Two warnings must be given, though. The first is that the thermal infrared signal will be characteristic of the surface only to a depth of the order of the attenuation length. This is only about 0.1 mm (see chapter 3), and oceanographers usually wish to know the temperature at greater depths than this. There is evidence that the temperature a few centimetres below the surface can be as much as 0.5 K warmer than at the surface (Robinson 1985).

The second warning, which applies to all calibrated thermal infrared observations, is that atmospheric attenuation, not entirely negligible in this waveband, will modify the signal somewhat, as will reflexion of

Fig. 5.6. Sea surface temperatures near northern Britain. The contours (in degrees Celsius) have been derived from thermal infrared data obtained on 17 May 1980 by the Advanced Very High Resolution Radiometer (AVHRR) carried aboard a TIROS satellite. The image was obtained from a height of approximately 820 km and has a coverage of about 800 × 900 km. (Reproduced by courtesy of Professor A.P. Cracknell, University of Dundee.)

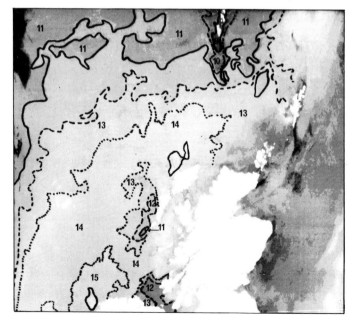

radiation from the sky. (Note also that the interpretation of thermal infrared imagery is often complicated by the presence of streaks and smears. These are temperature striations, caused by wind.) We will defer a discussion of the nature of this modification caused by atmospheric absorption and emission to chapter 6, since the effects on passive microwave observations are potentially more significant. For the present, we will note that the effect on a spaceborne observation is usually between 5 and 10 K, depending mainly on the amount of water vapour present in the atmosphere and on the presence of undetected clouds. This is clearly too large to ignore, and needs to be corrected using as much information as is available about the local atmospheric conditions. If two spectral channels are available within the thermal infrared band, a linear relationship of the form

$$\text{SST (true)} = aT_1 + bT_2 + c$$

is usually reliable to about 0.5 K (e.g. Robinson *et al.* 1984). In this expression, T_1 and T_2 are the brightness temperatures measured at two wavelengths (e.g. 11 and 12 μm, or 11 and 3.7 μm), and a, b and c are empirically determined constants. Because of the contribution from reflected sky radiation, the 'constants' have different values during the day and during the night. Grassl & Koepke (1981) give a comprehensive treatment of atmospheric corrections to SST measurements.

5.7.2 *Thermal inertia*

If the surface of a semi-infinite homogeneous slab of material is subjected to a periodically varying flux of thermal radiation, an attenuated wave of temperature variation will propagate into the slab. Two quantities are of particular interest: the attenuation depth (i.e. the depth at which the fluctuations are substantially less than at the surface) and the amplitude of the induced temperature fluctuations at the surface. The latter is determined largely by a quantity called the *thermal inertia*, which is characteristic of a given material and may sometimes be used to identify it.

The flow of heat inside a material is governed by the (thermal) *diffusion equation*,

$$\nabla^2 T = (\rho c / K) \partial T / \partial t \tag{5.6}$$

In this expression, T is the temperature above some arbitrary value, and K, c and ρ are the thermal conductivity, specific heat capacity and density of the material. This particular combination of variables, or rather its reciprocal, is, by analogy with other diffusion coefficients, termed the *diffusivity* Γ:

$$\Gamma = K/\rho c \qquad (5.7)$$

The general solution of (5.6) has been the subject of much work (see, for example, Carslaw & Jaeger 1959), but we can greatly simplify the problem by making a number of assumptions:

  (i) the material is homogeneous

  (ii) the incident flux is uniform over the surface, so that heat propagates only vertically (in the z-direction)

  (iii) the incident flux, and therefore the entire solution, varies sinusoid-ally in time with angular frequency ω

In this case it may easily be shown (e.g. by separation of the variables) that the only physically plausible solution of (5.6) is

$$T(z,t) = A \exp(i[z/z_0 - \omega t]) \exp(-z/z_0) \qquad (5.8)$$

where $z_0 = (2\Gamma/\omega)^{\frac{1}{2}}$ and A is a constant. This represents an exponentially decaying wave with an attenuation depth (to $1/e$) of z_0. If we differentiate (5.8) with respect to z we find that the temperature gradient at the surface ($z=0$) is

$$- A(\omega/\Gamma)^{\frac{1}{2}} \exp(-i\omega t - i\pi/4)$$

and since the flux F into the surface is given by $-K\partial T/\partial z$ at $z=0$, we have

$$F = KA(\omega/\Gamma)^{\frac{1}{2}} \exp(-i\omega t - i\pi/4) \qquad (5.9)$$

Comparing (5.8) and (5.9), we see that the flux and surface temperature variations are out of phase by $\pi/4$, in the sense that the flux is maximum $1/8$ cycle earlier than the temperature, and that the amplitudes T_0 and F_0 of these variations are related by

$$F_0/T_0 = K(\omega/\Gamma)^{\frac{1}{2}} = P\omega^{\frac{1}{2}} \qquad (5.10)$$

where P is the thermal inertia, defined as

$$P = (K\rho c)^{\frac{1}{2}} \qquad (5.11)$$

The most important periodicity for thermal inertia remote sensing is the daily cycle of solar heating ($\omega = 2\pi/86400\,\text{s}^{-1}$). The simple model which we have derived here is not strictly applicable, because the incident flux does not usually vary sinusoidally. The calculation of the flux is a complicated one, which includes contributions from direct solar radi-ation, radiation from the sky, geothermal heat flux and re-radiation from the surface itself (see Elachi 1987). It thus depends on whether the sky is clear or overcast, the geographical latitude and time of year, the emissivity of the surface, and other variable factors. Nevertheless, the model expressed by (5.8) gives a good indication of the trends which occur. Fig. 5.7 shows the typical variation of surface temperature with time of day, for materials of various thermal inertia. We should expect from (5.9) that it would reach a maximum value at 3 p.m. and a minimum at 3 a.m., whereas examination of fig. 5.7 suggests that the times of extreme temperature are in fact about 2 p.m. and just before dawn.

If surface temperatures are measured at (say) 2 a.m. and 2 p.m., differences in thermal inertia can be determined and, with accurate modelling, calculated. Table 5.1 lists the thermal properties of various materials, and they are illustrated in fig. 5.8. It can be seen that most minerals have values of P between about 1000 and 3000 $Jm^{-2}s^{-\frac{1}{2}}K^{-1}$, whereas (for example) metals have values about ten times larger, and wood about ten times smaller. There is little contrast between the thermal inertia of water and that of typical minerals. These materials may nevertheless be distinguished, since a water surface will normally be cooler than a rock surface during the daytime, as a consequence of evaporation. At night, the water surface is correspondingly warmer. A similar effect operates in the case of damp ground, and in this way soil moisture can be assessed. Dry vegetation may also be distinguished from

Fig. 5.7. Typical diurnal variations of surface temperature. The graphs show ΔT, the surface temperature minus the diurnal mean temperature, plotted against local solar time, for materials with thermal inertias of 400 and 2000 $Jm^{-2}s^{-1/2}K^{-1}$.

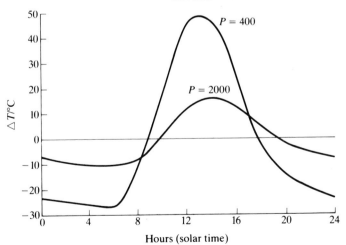

Table 5.1. *Thermal properties of various materials*

(In most cases, typical values are shown.)
The table gives values of the conductivity K, specific heat capacity c, density ρ, diffusivity Γ and thermal inertia P for various materials.

Material	K $\mathrm{W\,m^{-1}\,K^{-1}}$	c $1000\,\mathrm{J\,kg^{-1}\,K^{-1}}$	ρ $1000\,\mathrm{kg\,m^{-3}}$	Γ $10^{-6}\,\mathrm{m^2\,s^{-1}}$	P $1000\,\mathrm{J\,s^{-\frac{1}{2}}\,m^{-2}\,K^{-1}}$
basalt	2.1	0.9	2.6	0.9	2.2
clay (moist)	1.3	1.5	1.7	0.5	1.8
concrete	0.1	3.4	2.4	0.01	0.9
copper	400	0.39	8.9	120	37
dolomite	2	0.75	2.6	1.0	2.0
glass	0.8	0.6	2.3	0.6	1.1
granite	3	0.8	2.7	1.4	2.5
gravel	1.3	0.8	2.0	0.8	1.4
gravel (sandy)	2.5	0.8	2.1	1.5	2.0
ice	2.3	2.1	0.9	1.2	2.1
iron	84	0.44	7.9	24	17
lead	36	0.16	11	20	8.0
limestone	0.9	0.7	2.5	0.5	1.3
marble	2.5	0.9	2.7	1.0	2.5
pumice (loose)	0.3	0.7	1.0	0.4	0.5
quartz	9	0.7	2.6	4.9	4.0
sand (dry)	0.4	0.8	1.6	0.3	0.7
shale	1.9	0.7	2.3	1.2	1.7
soil (sandy)	0.6	1.0	1.8	0.3	1.0
steel	25	0.5	7.7	6.5	9.8
water	0.56	4.2	1.0	0.13	1.5
wood	0.15	1.2	0.7	0.2	0.4

bare ground at night, since the vegetation and the ground beneath it is physically warmer than the bare ground, because of the insulating effect of the vegetation. A converse effect operates in the daytime.

Thermal inertia mapping was carried out from space, during the HCMM (heat capacity mapping mission). This satellite carried a thermal infrared radiometer (10.5–$12.5\,\mu m$ band) called the HRIR (high resolution infrared radiometer), which had an accuracy of $0.4\,K$. The instrument in fact also observed in other spectral bands, measuring atmospheric temperature, pressure and humidity profiles, and also cloud, water vapour and carbon dioxide concentrations, with channels between 4 and $8\,\mu m$ and between 13 and $15\,\mu m$, and ozone concentration with a channel at $9.7\,\mu m$. The results from the HCMM indicate that thermal inertia mapping is particularly sensitive to the effects of tectonic disturbance and to lithological boundaries. However, distinguishing one rock type from another is still rather difficult. Soil moisture, on the other hand, can be estimated to an accuracy of about 15% using thermal inertia measurements, and the technique has also found application to vegetation monitoring, and indeed to the study of volcanoes. Plate 5 shows an example of the type of image generated by the HCMM.

Thermal inertia mapping also has an application to archaeological surveying. If one material is buried within another, and the materials have differing thermal properties, the flow of heat will be distorted and give rise to a *temperature anomaly* at the surface. This will yield information on the nature of the buried object, but the interpretation is complicated by the fact that the simple one-dimensional problem which we have been describing is no longer applicable, and the diffusion equation (5.5) must be solved in three dimensions. Tabbagh (1973) discusses possible applications of the technique.

Fig. 5.8. Thermal diffusivities (Γ) and inertias (P) of various materials. The diagonal lines are labelled with the corresponding values of the volumetric heat capacity in $Jm^{-3}K^{-1}$.

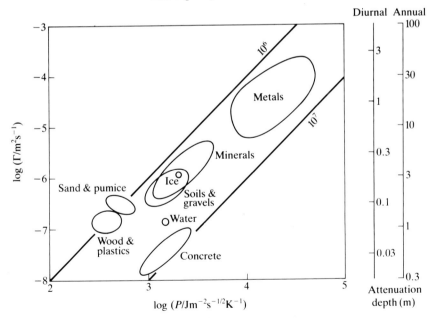

Problems

1. Show that the resolution limits imposed by diffraction and by finite detector size are approximately equal if the f/number of the focussing device is equal to the ratio of the detector element size to the wavelength.

2. Evaluate the sensitivity S of the Planck radiation function at a temperature of 300 K and a wavelength of (a) $4\,\mu m$, (b) $11\,\mu m$.

3. Show that the ratio of amplitudes of the net flux and surface temperature variations for a uniform homogeneous material subjected to a sinusoidally varying heat input is $P\omega^{\frac{1}{2}}$, where P is the thermal inertia and ω is the angular frequency. For this idealised case, show also that such a surface should reach its minimum diurnal temperature at about 3 a.m. At what date will the annual component be a minimum?

4. Include the effect of incident solar radiation within the model of question 3, by assuming that the incident solar flux varies sinusoidally with an amplitude F_s, and that the flux reradiated upwards from the surface is equal to a constant c times the surface temperature. Show that the amplitude of the surface temperature fluctuations is given by $T_0 = F_s(P^2\omega + Pc[2\omega]^{\frac{1}{2}} + c^2)^{-\frac{1}{2}}$. Assuming that $F_s = 200\,\mathrm{Wm}^{-2}$ and $c = 3\,\mathrm{Wm}^{-2}\mathrm{K}^{-1}$, calculate T_0 for daily fluctuations when $P = 400$ and $2000\,\mathrm{Jm}^{-2}\mathrm{s}^{-\frac{1}{2}}\mathrm{K}^{-1}$. Approximately what emissivity is represented by this value of c?

6

Passive microwave systems

6.1 Introduction

In chapters 4 and 5 we considered passive remote sensing systems in which the diffraction resolution limit λ/D, while important, was not usually a critical parameter of the operation. In this chapter, we shall consider our last type of passive remote sensing system, the passive microwave radiometer. This is a device which measures thermally generated radiation in the microwave (usually 5–100 GHz) region. As we discussed in section 2.6, the long 'tail' to the Planck distribution at relatively low frequencies means that significant amounts of radiation are emitted even in this range of frequencies.

Because microwave wavelengths are so much greater than those of visible or even infrared radiation, the resolution limit plays a much more important role, and we shall need to give more careful consideration to the factors which determine it. The treatment which follows in this chapter is similar to that of Robinson (1985), and is expanded upon in Ulaby *et al.* (1981). Much of the technology and nomenclature of passive microwave radiometry was originally developed in the field of radio astronomy, and further details can also be found in works on this subject.

6.2 Antenna theory

As we have remarked before, electromagnetic radiation is detected by means of its influence on electrons, which are excited to higher energy states by the incident photons. The energy of a microwave photon is typically only a few μeV, which is too small to excite an electron across an atomic or molecular band-gap. For this reason, *conductors* (metals) are used. The incident electromagnetic wave induces a fluctuating current in the conductor, which can subsequently be amplified and detected. The *antenna* is a structure which serves as a transition between the wave propagating in free space, and the fluctuating voltages in the circuit to which it is connected.

The usual form of a microwave antenna is a paraboloidal dish, although many other designs are possible. Fig. 6.1 shows schematically the design of a simple microwave radiometer using such an antenna.

The design of an antenna is dominated by two considerations:
 (i) the need to achieve a high sensitivity in the desired direction
 (ii) the need to achieve a high angular resolution (narrow beamwidth)
Both of these requirements are met by making the antenna as large as possible.

Let us consider a perfect antenna, in which there are no power losses. It has a *radiation resistance R*, a fictitious quantity which can most easily be understood in the case of a transmitting antenna. In this case, R represents the element of the circuit in which power is dissipated, although of course in reality no dissipation takes place and the power is transmitted in the form of an electromagnetic wave.

If the antenna (now in receiving mode again) is directed at a large, distant region emitting microwave energy, a voltage signal will appear at the output of the antenna. If the emission mechanism is thermal, that is the signal is noise-like, it will have exactly the same characteristics as the *thermal noise* generated in a resistance R held at a temperature T_A. The latter is in essence caused by the Brownian motion of electrons in the resistance, and is often referred to as Johnson noise or Nyquist noise. The magnitude of this term is derived (see e.g. Bleaney & Bleaney, 1976;

Fig. 6.1. Construction of a passive microwave radiometer (schematic). The antenna collects the incident radiation and generates a fluctuating voltage which can then be amplified and processed.

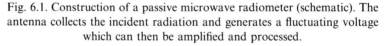
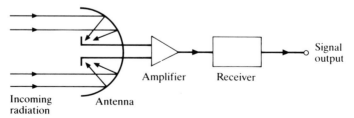

Longair, 1984) using a combination of electrical and statistical mechanical arguments, and found to be

$$P_N = kT_A \Delta v \qquad (6.1)$$

where k is Boltzmann's constant and Δv is the signal bandwidth. (Note that thermal noise thus has constant power per unit frequency interval.) T_A is then called the *antenna temperature*. If the distant region is large enough, and has the radiation characteristics of a black body, the antenna temperature will be equal to the physical temperature of the target.

The ideal antenna would receive radiation uniformly over a small range of solid angle. To describe a real antenna, we introduce the idea of the *power pattern* $P(\theta, \phi)$. This is the power which would be detected by the antenna from a point transmitter of fixed strength, located in the direction (θ, ϕ) with respect to the antenna axis, but at a fixed distance greater than the Fresnel distance (see chapter 2). The power pattern is usually normalised, so that

$$P_n(\theta, \phi) = P(\theta, \phi) / P_{max}(\theta, \phi)$$

Fig. 6.2 illustrates a typical power pattern (or, at least, its θ variation). It has a *main lobe* of maximum sensitivity, usually in the direction $(0,0)$, and a number of undesirable *sidelobes*. The usual measure of the width of the main lobe is its half-power beam width (HPBW), defined as the angular width of the region in which $P_n \geq 1/2$. For antennas with large

Fig. 6.2. Power pattern of a typical antenna, plotted in polar coordinates. The half-power beam width (HPBW) is defined as the angular width of the region in which $P_n(\theta) > 1/2$.

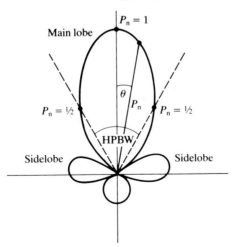

Table 6.1. *Properties of various antenna types*

The table shows the HPBW, directivity D and maximum sidelobe level for a few of the most important types of antenna. In the last two types, it is assumed that the physical dimensions a, b and D of the antennas are much greater than the wavelength of the radiation.

Antenna type	HPBW (degrees)	D (decibels)	Sidelobes (decibels)	Notes
monopole	—	0	(none)	isotropic
short dipole	90	1.76	(none)	isotropic in perpendicular direction
half-wave dipole	90	2.15	(none)	isotropic in perpendicular direction
6-element Yagi	42	10.5	-10	TV antenna-type
rectangular	$51(\lambda/a)$	$11 + 10\log_{10}(ab/\lambda^2)$	-13	a, b are sides of rectangle
circular paraboloid	$72(\lambda/D)$	$\approx 10 + 20\log_{10}(D/\lambda)$	-25	D is diameter

apertures, the power pattern is calculated by Fourier Transform methods, since it is effectively the square of the diffraction pattern of the aperture distribution. For antennas which are smaller than a few wavelengths in any direction perpendicular to that of the incident radiation, the calculation is more complicated and requires the application of electrodynamics. Table 6.1 lists the HPBWs of some common designs of antenna. The normal measure of the strength of the sidelobes is the peak value of P_n, expressed in dB. Thus it may be said that a well designed antenna will have sidelobe levels of -20 dB or lower. The *beam solid angle* is defined as

$$\Omega_A = \int_{4\pi} P_n(\theta,\phi)\,d\Omega \tag{6.2}$$

where $d\Omega = \sin\theta\,d\theta\,d\phi$, and the *main beam solid angle* is similarly defined as

$$\Omega_M = \int_{\text{main lobe}} P_n(\theta,\phi)\,d\Omega \tag{6.3}$$

We saw in chapter 2 that the spectral radiance of a black body at a temperature T is given by

$$L_v = \frac{2hv^3}{c^2}\frac{1}{e^{hv/kT}-1} \tag{2.43}$$

If $h\nu/kT \ll 1$, which will be true for passive microwave radiometry of objects at ordinary terrestrial temperatures, this can be approximated by

$$L_\nu = 2kT/\lambda^2 \qquad (6.4)$$

which is called the *Rayleigh–Jeans approximation*. (By comparison with (2.43), it can be shown that this expression is accurate to better than 1% for objects at 300 K if the frequency is below about 125 GHz.) Equation (6.4) defines the *brightness temperature* of an object in terms of L_ν. If the object is a black body, its brightness temperature will be equal to its physical temperature. We can also see from the form of (6.4) that the brightness temperature T_b is in general given by

$$T_b = \varepsilon T_p \qquad (6.5)$$

where ε is the emissivity and T_p the physical temperature. This expression is only valid if (a) the emission mechanism is thermal and (b) the Rayleigh–Jeans approximation is valid.

Thus if we have a uniform source subtending a solid angle $d\Omega_S$, the *spectral flux density* reaching the antenna from the source is

$$F_\nu = 2kT_b d\Omega_S/\lambda^2 \ (\mathrm{Wm^{-2}Hz^{-1}}) \qquad (6.6)$$

If the source contains a distribution of brightness temperatures, the antenna will observe an average flux density equal to

$$F_\nu = \frac{2k}{\lambda^2} \int_{4\pi} T_b(\theta,\phi) P_n(\theta,\phi) \, d\Omega$$

We can clearly define an *effective area* A_e of the antenna, such that the power collected per unit bandwidth is $SA_e/2$ (the factor of 1/2 arises because we are assuming that the antenna receives only one polarisation – see chapter 2). This will also, by (6.1), be equal to kT_A so that we have

$$T_A = \frac{A_e}{\lambda^2} \int_{4\pi} T_b(\theta,\phi) P_n(\theta,\phi) \, d\Omega \qquad (6.7)$$

but it is also clear that T_A will be a weighted average of the brightness temperatures seen by the antenna:

$$T_A = \frac{\displaystyle\int_{4\pi} T_b(\theta,\phi) P_n(\theta,\phi) \, d\Omega}{\displaystyle\int_{4\pi} P_n(\theta,\phi) \, d\Omega}$$

$$= \frac{\displaystyle\int_{4\pi} T_b(\theta,\phi) P_n(\theta,\phi) \, d\Omega}{\Omega_A}$$

Comparing this expression with (6.7) shows that the effective area and the beam solid angle are related by

$$\Omega_A A_e = \lambda^2 \tag{6.8}$$

In the case of large (much greater than the wavelength λ in directions perpendicular to the radiation) antennas, A_e will be approximately equal to the geometrical area of the antenna. This can be seen roughly by noting the equivalence of P_n to the diffraction pattern of the antenna, noted earlier, and (2.32).

Let us apply this theoretical development to two examples, representing opposite extremes. Firstly, we shall take a target object with a brightness temperature T_b and an angular size $\Delta\Omega_S$ much smaller than the beam solid angle Ω_A. From (6.7) and (6.8) it is clear that the antenna temperature will be $T_b \Delta\Omega_S/\Omega_A$, and a larger antenna (which will have a smaller beam solid angle) will give a larger signal. However, if the antenna is large enough that the beam solid angle Ω_A is much smaller than the angular size of the target (i.e. the target is resolved), the antenna temperature will be T_b. A further increase in the size of the antenna will *not* increase the power detected by it, but will of course improve the resolution. This at first sight paradoxical result is explained by realising that although a larger antenna collects more radiation, it also collects this radiation from a smaller region of the target. The two effects cancel one another out.

There are two further terms, useful in describing the properties of an antenna, which we ought to introduce. The *directivity D* is defined as

$$D = \frac{4\pi}{\int_{4\pi} P_n(\theta,\phi)\, d\Omega}$$

from which it follows that

$$D = 4\pi A_e/\lambda^2 \tag{6.9}$$

The directivity, which is often quoted in dB, is a measure of the sensitivity of the antenna in its most sensitive direction. If θ_0 and ϕ_0 are the HPBWs in orthogonal directions, expressed in *degrees*, the directivity is given approximately by

$$D \approx 3 \times 10^4/\theta_0\phi_0 \tag{6.10}$$

Up to this point, we have assumed that the antenna is lossless. Usually, the properties of an antenna are expressed in terms of its *gain G*, which includes the effects of ohmic (resistive) losses. If we write A_e' for the effective area taking these losses into account, we have

$$G = 4\pi A_e'/\lambda^2 \tag{6.11}$$

The gain is often expressed in dB relative to that of a monopole or dipole antenna.

6.3 Sensitivity

We have noted that it is usual to express the power per unit bandwidth extracted by an antenna as a temperature (the antenna temperature). This facilitates calculation of the *sensitivity* of the system, since the noise power generated by the system itself is also normally expressed as a temperature. The system noise temperature depends on the detailed design of the receiver and other parts of the system, but it cannot be lower than the physical temperature of the receiver, and will usually (at the frequencies typical of passive microwave radiometry and for a well designed receiver) be a factor of 1.2–2 times this value.

In order to improve its signal-to-noise ratio, the output from a radiometer is integrated (averaged) for a time Δt. If the bandwidth over which the radiation is received is Δv, we may regard this as an average over $N = \Delta t \Delta v$ independent samples. The signal-to-noise ratio will thus be improved by a factor of $N^{\frac{1}{2}}$, and we expect the sensitivity of the system to be

$$\Delta T = C T_{\text{sys}} (\Delta t \Delta v)^{-\frac{1}{2}} \tag{6.12}$$

In this expression, ΔT is the smallest change in antenna temperature which can be detected, T_{sys} is the system noise temperature, and C is a factor of order unity which depends on both the precise design of the radiometer and the criterion used to define the smallest detectable change. C usually has a value of 5–10.

6.4 Scanning radiometers

A microwave radiometer has a spatial sensitivity defined, as we have seen, by the power pattern $P_n(\theta, \phi)$. This corresponds to a region of sensitivity, usually termed the *footprint* (although the concept is evidently equivalent to the IFOV), at the earth's surface. It is clearly desirable to be able to scan this footprint from side to side in order to acquire information from a wide swath, and this can be achieved in two ways.

The obvious method of scanning a radiometer footprint is by mechanical steering of the antenna. The antenna itself may rotate (or oscillate) with respect to the rest of the instrument, although this can cause some mechanical difficulties, or the whole platform can be made to rotate. The latter approach is clearly more appropriate for spaceborne systems.

Any form of mechanical scanning is likely to cause some residual, undesirable, oscillation or vibration of the instrument (especially in view of the fact that passive microwave systems usually have large and therefore heavy antennas in order to achieve reasonably small footprints), leading to pointing errors and loss of resolution. Another approach, which avoids these problems, is provided by electronic scanning.

An electrically-scanned antenna has no moving parts. It consists of a closely-spaced regular array of smaller antennas (e.g. waveguides, horns or dipoles). The signals detected at each of these elements can be advanced or retarded in phase under electronic control (hence the alternative name of *phased array* for this type of system), and in this way steering is achieved.

Fig. 6.3 illustrates the principle of electrical steering in one dimension. It shows a simplified array, having only eight detectors at spacing d. The signals from these detectors are shifted in phase and then added together. If no phase shift is applied to any of the signals, and the detectors abut each other so that there are no 'holes' in the composite antenna, the device clearly functions as an ordinary antenna of width $8d$, having a power pattern with a main lobe pointing in the direction $\theta = 0$ and a width of approximately $\lambda/8d$ radians.

If, however, we consider radiation arriving at an angle θ to the normal, it is clear that the phase of the signal received at detector 2 is $kd\sin\theta$ in advance of that at 1 (k being the wavenumber $2\pi/\lambda$), the phase at 3 is in advance of that at 2 by the same amount, and so on. In order to add these signals constructively, we arrange that the phase shifters introduce phase delays ϕ_1 to ϕ_8 which will exactly compensate for this. Thus by introducing a phase gradient across the array, we can shift the direction of its main lobe.

Electronic steering of an array naturally has its disadvantages. One is the increased complexity of the system, since a practical system will contain many more than eight elements, and probably be steerable in two dimensions instead of just one. A second drawback is the decrease in performance caused by the inevitable errors in setting the phases and gains of the phase shifters. However, a problem which is fundamental rather than technological is provided by the fact that the *projected* length of the antenna perpendicular to the incoming radiation is proportional to $\cos\theta$, so that as the scan angle (away from normal) is increased, the HPBW will widen accordingly. This can be an important consideration if an instrument is required with a very wide swath.

6.5 Atmospheric effects

As we saw in chapter 3, the atmosphere is not completely transparent in the microwave region of the electromagnetic spectrum. The attenuation increases with frequency from 1 to about 1000 GHz, and there are also several absorption bands. Typically, the attenuation can be crudely characterised by a uniform attenuation coefficient from sea level up to some characteristic scale height (usually 2–3 km), and zero attenuation above that. The approximate value of the total atmospheric attentuation, as a function of frequency, was shown in fig. 3.10 for a ray propagating vertically. For a ray making an angle θ to the vertical, the attenuation in

Fig. 6.3. An electrically-steered (phased) array. The signals from the elements 1–8 are given phase shifts ϕ_1 to ϕ_8 so that radiation arriving from the direction θ is combined in phase.

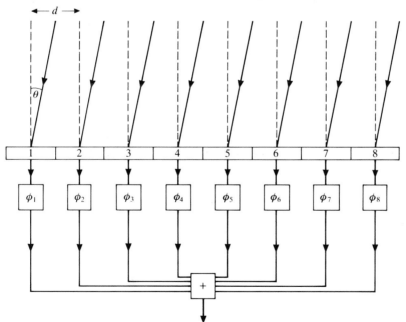

decibels may be assumed to be inversely proportional to $\cos\theta$ (i.e. proportional to the path length through the atmosphere) for angles up to about 80°.

The effect of atmospheric absorption is somewhat beyond our scope to allow for in the most general case, but a simplified model, treating the case of plane parallel radiation in a uniform medium, gives a good understanding of what happens.

If a signal corresponding to a brightness temperature T_0 passes through a channel (defined in the broadest sense to include attenuators as well as atmospheric paths) whose attenuation is A (i.e. the signal is multiplied by a factor A which is less than unity) and whose physical temperature is T_a, the observed temperature T_1 will be the sum of two parts:

(i) the origina' signal, but attenuated
(ii) a contribution from the channel itself

It is clear that the first part of the signal must be equal to $T_0 A$, and it follows from a simple thermodynamic argument that the second part is $T_a(1-A)$. Thus

$$T_1 = T_0 A + T_a(1-A) \tag{6.13}$$

This is in fact a very simple case of the *radiative transfer equation*. As an example, let us assume that a region of the earth's surface, whose brightness temperature is 200 K, is being observed at a frequency at which the total atmospheric attenuation is 0.5 dB. Thus $A = 10^{-0.05} = 0.891$, and if $T_a = 280$ K, T_1 will be 208.7 K. In this example we may describe the effect contributed by the atmosphere as an *upwelling brightness temperature* of 30.4 K, and it can be seen by inspection of fig. 3.10 that the vertically upwelling brightness temperature will range from about 4 K to about 200 K over the range of frequencies from 1 to 50 GHz.

If the effect of the atmosphere is to contribute an upwelling signal at the top of the atmosphere, it must also produce a *downwelling brightness temperature* at the surface, which can be calculated in the same way as the

Fig. 6.4. A signal corresponding to a brightness temperature T_0 passes through a channel which multiplies its power by A (less than unity). The temperature of the channel is T_a and the brightness temperature of the output signal is T_1.

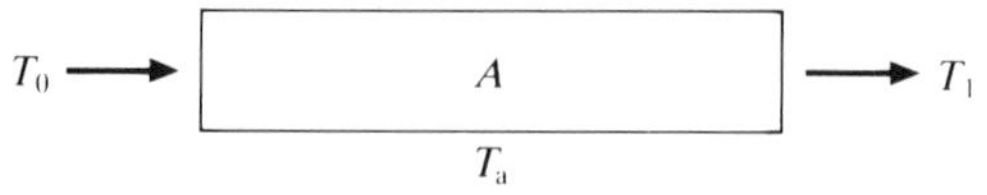

$$T_0 \longrightarrow \boxed{\qquad\qquad A \qquad\qquad} \longrightarrow T_1$$
$$T_a$$

upwelling signal. If the surface from which the radiometer is detecting radiation has a low emissivity (and consequently a high reflectivity), the possibility exists that a significant amount of this downwelling radiation may be reflected by the surface into the radiometer. This effect is generally insignificant except for horizontally polarised radiation at very large incidence angles, where the emissivity is small (see section 6.7). These effects (which are summarised diagrammatically in fig. 6.5), though significant, can be corrected fairly easily using a model of the atmosphere or, in critical cases, *in situ* sounding.

We should also mention the effects of fog, cloud and rain on microwave radiation. In the case of fog and cloud, the droplets are smaller than about 0.1 mm and the attenuation is mostly caused by absorption rather than scattering. An approximate model for the attenuation coefficient is given by (6.14), valid for frequencies between about 1 and 100 GHz (Ulaby *et al.* 1981).

$$\alpha = 0.6(v/\text{GHz})^{1.9}(\rho/\text{kg m}^{-3})\,\text{dB/km} \tag{6.14}$$

In this expression, ρ is the mass density of water in the cloud. Typical values of this parameter range from $10^{-6}\,\text{kg m}^{-3}$ for haze, to $10^{-4}\,\text{kg m}^{-3}$ for fog, to about $10^{-2}\,\text{kg m}^{-3}$ for cumulonimbus cloud. The corresponding meteorological visibility for an optically thick layer of fog or cloud (defined as the distance over which visible-wavelength radiation suffers an attenuation of 18 dB) is given approximately by

$$\text{visibility (m)} \approx 0.4(\rho/\text{kg m}^{-3})^{-0.7}$$

The typical layer thicknesses range from 50 m (fog) to of the order of 1000 m (haze and cloud).

Fig. 6.5. Contributions to a passive microwave observation of the earth's surface. The diagram attempts to show the attenuation of the signals from the surface, but otherwise no significance should be attached to the sizes of the arrows.

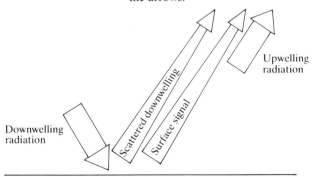

The effect of rain depends principally on the rate R of rainfall. An approximate, empirical expression for the attenuation coefficient due to rain is given by (6.15)

$$\log \alpha = -4.4 + 2.42 \log v + 1.41\, v^{-0.078} \log R \qquad (6.15)$$

In this expression, α is measured in dB/km, v in GHz and R in mm/hour. The logarithms are to base 10. Typical values for R range from 0.25 mm/hr in a light drizzle, 2.5 in light rain, 25 in heavy rain to 100 mm/hr in a tropical downpour. The thickness of a rain layer is typically 4000 m.

The microwave attenuation coefficient of falling snow is such that $\log \alpha$ is smaller than the corresponding value for rain by typically 1–2, i.e. the value of α is smaller by a factor of 10–100 (Veck 1985).

6.6 Design of a radiometer – the SMMR

The SEASAT satellite carried, amongst other instruments, a passive microwave radiometer called the SMMR (scanning multichannel microwave radiometer). Although SEASAT survived for only three months after its launch in June 1978, a SMMR is still in operation on the NIMBUS series of satellites.

The SMMR, shown schematically in fig. 6.6, has an offset paraboloidal mirror with a diameter of 0.79 m. Because of its inclination, it presents to

Fig. 6.6. Operation of the scanning multichannel microwave radiometer (SMMR) carried by the SEASAT and NIMBUS satellites.

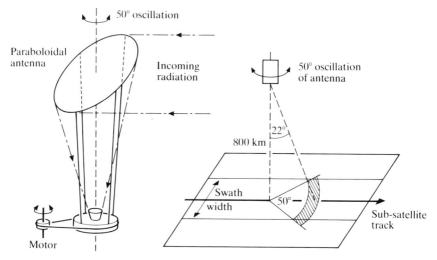

the incoming radiation an elliptical aperture of dimensions 0.79×0.52 m, and in consequence the footprint upon the earth's surface is also elliptical. The radiation is reflected into a horn, from which it is transmitted via a waveguide to receivers tuned to frequencies of 6.63, 10.6, 18.0, 21.0 and 37.0 GHz. The bandwidth of these receivers are all 220 MHz, and the integration times are 126, 62, 62, 62 and 30 ms respectively.

From table 6.1 we see that the HPBW of a paraboloidal antenna is approximately $72 \lambda/D$ degrees, so we expect the beamwidth of the 10.6 GHz channel, for example, to be about $2.6° \times 3.9°$. At a height of 800 km and an incidence angle of 22°, this corresponds to a footprint of about 40×60 km. We would expect the sensitivity, using (6.12), to be a few times 0.1 K, and in fact the observed value is 0.9 K. It is interesting to note, using (6.1), that this corresponds to a power sensitivity of only 3×10^{-15} W.

6.7 Applications of passive microwave radiometry

We have seen how a passive microwave radiometer generates a signal which depends on the brightness temperature of the target, and that the brightness temperature T_b is related to the physical temperature T_p by (6.5).

A given measurement may be regarded as either a determination of T_p if ε is known, or a determination of ε if T_p is known. In the former mode it could be used, for example, to determine the surface temperature of the sea, and in the latter it finds widespread application for determining type and concentration of sea ice.

The variation of ε with frequency and polarisation shown by many materials allows the use of multiparameter observations analogous to the multispectral approach discussed in chapters 4 and 5, and again in chapter 10. For radiometer observations made at angles away from the surface normal, differences in emissivity for horizontally and vertically polarised radiation are predicted by equations (2.38) for simple homogeneous materials. Fig. 6.7 shows, for example, the emissivity of a water surface as a function of incidence angle.

6.7.1 Oceanographic applications

The main use of passive microwave radiometer data to date has been to measure SST, which can be done with an absolute accuracy of about 1 K and a relative accuracy of about 0.2 K (see plate 6). Since the brightness temperature is influenced not only by the physical temperature (the SST) but also by the observation frequency, the salinity and the state of the surface (because of their effects on the emissivity), an accurate determination of the SST requires a multifrequency observation. The attenuation length for microwaves in seawater is of the order of 1 cm, that is, considerably greater than in the thermal infrared band. A comparison of SSTs derived using the two techniques thus has scope for investigating the surface temperature anomaly described in section 5.7.1.

Passive microwave radiometry has been used over ocean surfaces to determine *wind speed*, with an accuracy of about $2\,\mathrm{ms}^{-1}$. This method relies on the influence of wind speed upon surface roughness, and hence upon the emissivity. *Salinities* can also be determined through their effect on the emissivity, but (so far) only from aircraft. The reason for this is that sensitivity to salinity occurs only at frequencies below about 2 GHz, and an antenna operating at such a low frequency would need to be very large to achieve a useful spatial resolution.

Passive microwave radiometry has proved enormously useful in the delineation of sea ice, since seawater and ice have substantially different

Fig. 6.7. Emissivity of a pure water surface at 10 GHz and 100 GHz, showing both vertical and horizontal polarisations, as a function of the angle θ between the emitted radiation and the surface normal.

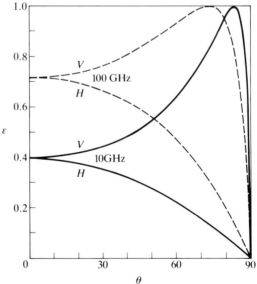

emissivities. Single-frequency observations allow the fractional ice cover to be determined by interpolation, and multifrequency and/or multipolarisation observations can allow different ice types to be distinguished (Carsey 1985, Parkinson *et al.* 1987). Plate 7 shows an example of this.

6.7.2 *Meteorological applications*

Passive microwave radiometry can measure a number of useful meteorological parameters. We have already mentioned SST and wind speed over oceans, but if the instrument is tuned to one of the atmospheric absorption bands water vapour content, cloud liquid water content, rainfall rate and atmospheric temperature profiles can also be determined.

6.7.3 *Land applications*

It is much harder to use passive microwave systems over land surfaces, except from low-altitude airborne platforms. The reason is the large angular beamwidth of such systems, giving footprints as large as 100 km from satellites. Such large footprints generally contain a wide range of materials with different emissivities, and interpretation of the observations is very difficult. There are a few promising lines of research, though. The superficial extent and depth (to about 0.2 m – see for example Hallikainen 1984) of snowfields can often be determined, as can soil moisture content, below about 5 GHz. For large areas of uniform composition, such as deserts, surface temperatures can be found.

Problems

1. A terrestrial antenna transmits a signal of power 1 kW at 5 GHz to a geostationary satellite 35 800 km away. If the transmitting antenna has a diameter of 10 m and the receiving antenna has a diameter of 1 m, what is the received power? (Assume both antennas to have unit efficiency.)
2. At 10.4 GHz, the brightness temperatures of sea water, first year ice and multi-year ice are 80 K, 252 K and 200 K respectively. At 36 GHz these figures become 119 K, 253 K and 168 K. If a microwave radiometer measures a brightness temperature of 180 K at both frequencies, what are the fractions of open water and multi-year ice present in the IFOV?

3. Plot the variation in normal brightness temperature, over the range 1 to 100 GHz, of a water surface of physical temperature 20 °C as observed by a satellite radiometer. You may assume the atmosphere to have a uniform temperature of 280 K.

4. Repeat question 3 for the case when there is an intervening layer of cloud 2 km thick, with a water content of 10^{-3} kg m^{-3} and a temperature of 0 °C. Is the height of the cloud layer important?

5. Explain why the emissivity of sea water is less than that of pure water under the same conditions. Plot the variation in the normal emissivity of a water surface from 0.1 to 10 GHz. On the same graph, plot the emissivity of sea water, assumed to have a conductivity of 5 Sm^{-1}. At what frequency is the emissivity of sea water 10% less than that of pure water?

7

Ranging systems

7.1 Introduction

In chapters 4 to 6 we have considered *passive* sensors, detecting naturally occurring radiation. In this chapter and the next we shall consider *active* sensors, which themselves illuminate the earth's surface. We have already discussed the division of sensors into imaging and non-imaging types, and their classification by the wavelength of the radiation employed. There is however a third way in which we can classify active sensors, and that is according to the use which is made of the returning signal. If we are principally interested in the time delay between transmission and reception of a signal we shall call the method a *ranging technique*, whereas if we are mainly interested in the strength of the returning signal we shall call the method a *scattering technique*. The division between the two cannot be made entirely rigorous, but it provides a useful way of thinking about active remote sensing systems. It is clear that ranging systems are simpler both to visualise and, because of their less stringent technical requirements, to construct, and we shall therefore consider them first. In chapter 8 we will discuss the scattering techniques.

7.2 Laser profiling

Laser profiling is a ranging application of the LIDAR technique, LIDAR being an acronym formed in the same manner as RADAR and standing for LIght Detection And Ranging. We note in passing that the principal use of LIDAR is an an atmospheric sounding technique, and therefore beyond the scope of this book. Both ranging and sounding applications of LIDAR are discussed in detail by Measures (1984).

The principle of laser profiling is extremely simple. A pulse of 'light' (usually in fact infrared radiation) is emitted towards the surface by the instrument, and its 'echo' is detected some time later. By measuring the time delay and knowing the speed of propagation of the light pulse, the range (distance) from the instrument to the surface can be determined. By sending out a continuous stream of pulses, a profile of the range can be built up, and if the position of the platform as a function of time is accurately known, the surface profile may then be deduced.

The operation and construction of a typical laser profilometer are shown schematically in figs. 7.1 and 7.2. The transmitter is a gallium arsenide (GaAs) semiconductor laser, which is capable of producing a short, intense pulse with a small IFOV at a wavelength of about 0.9 μm. The receiver is a photodiode (see chapter 5). An interval timer with a resolution of the order of 1 ns is started by the signal which generates the transmitted light pulse, and stopped by the received pulse. The travel time of the pulse, T_t, is given by

$$T_t = 2H/c$$

where H is the range and c is the speed at which the pulse propagates.

The desirable features of such a system are that it should achieve a high spatial resolution at the surface (i.e. the sampled points should be close

Fig. 7.1. Principle of operation of a laser profiler.

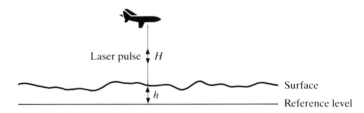

Fig. 7.2. Construction of a laser profiler (schematic).

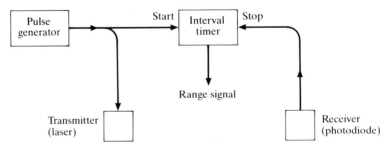

together) and a high range resolution, and that the sensitivity should be great enough to detect signals returning from weakly reflecting surfaces.

The accuracy ΔT_t with which the travel time can be determined is normally governed by the *rise time* t_r of the received pulse, and its signal-to-noise ratio S. This can be understood by referring to fig. 7.3. If V_s is the voltage amplitude of the received pulse and V_n is the amplitude of its variation due to noise, the (voltage) signal-to-noise ratio is defined as $S = V_s/V_n$. It is evident from the figure that the greatest accuracy with which the timing of the received pulse can be determined is given by

$$\Delta T_t = t_r/S \tag{7.1}$$

although the accuracy of the system may in fact be limited by that of the interval timer.

A typical system will have a rise time of a few nanoseconds for the transmitted pulse, although the received pulse may be somewhat longer if the surface being profiled is particularly rough (this is discussed in greater detail in section 7.3). The signal-to-noise ratio S will depend on the reflectivity of the surface and the range H, as well as on system parameters such as the transmitted power, and less easily calculated influences such as the amount of incident sunlight, the weather, and atmospheric attenuation.

Fig. 7.3. Determining the arrival time of a noisy pulse. The voltage rises during a time t_r from a mean value of zero to a mean value of V_s. If the noise level is V_n, the arrival time can be determined to an accuracy of $\Delta T_t = t_r V_n/V_s$.

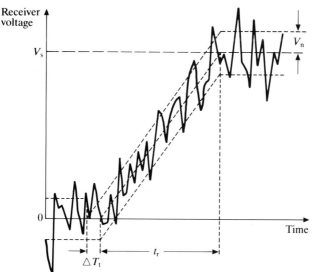

If pulses are transmitted at a frequency p (called the *pulse repetition frequency* or PRF) and the platform velocity is v, the linear sampling interval is v/p. If the angular IFOV of the system is $\Delta\theta$, the linear footprint will be $H\Delta\theta$. It is desirable that this should be no larger than some maximum value set by the nature of the surface and the type of investigation. It might be imagined, however, that there is no point in decreasing the value of v/p below $H\Delta\theta$, but this is not so. Any decrease below this value means, in effect, that a number of independent measurements of the range is being made over a single footprint, and this will improve the range accuracy. Since the number N of independent measurements is $pH\Delta\theta/v$ and the improvement in accuracy is proportional to $N^{\frac{1}{2}}$, we may write the range accuracy as

$$\Delta H = \frac{ct_r}{2S}\left(\frac{v}{pH\Delta\theta}\right)^{\frac{1}{2}} \tag{7.2}$$

A typical airborne system might have $t_r=5\,\mathrm{ns}$, $S=1$, $v=50\,\mathrm{ms}^{-1}$, $p=1000\,\mathrm{s}^{-1}$ and $H=200\,\mathrm{m}$. If the angular IFOV is 10^{-3} radians (a typical figure for a laser), the minimum linear footprint is $0.2\,\mathrm{m}$, $N=4$ and $\Delta H=0.4\,\mathrm{m}$. If the angular IFOV is artificially degraded (by averaging pulses) to 0.025 radians, the linear footprint becomes $5\,\mathrm{m}$, $N=100$ and $\Delta H=0.08\,\mathrm{m}$.

Equation (7.2) shows that the measurement accuracy is increased by increasing the PRF. However, if p is increased beyond a certain point, the measured range will become ambiguous, for if the travel time T_t exceeds the interpulse period $1/p$ it will not be certain which echo belongs to which transmitted pulse. The calculated range will suffer from a *range ambiguity* of

$$H_{amb}=c/2p \tag{7.3}$$

in the sense that the calculated range H may be increased or decreased by integral multiples of H_{amb} without changing the apparent travel time. This is an example of the aliasing phenomenon which is discussed in greater detail in chapter 9. Airborne laser profilometers are normally operated at a PRF such that $H_{amb}\gg H$, so that the echo of a given pulse returns well before the next pulse is emitted. No ambiguity therefore arises.

Laser profiling has found applications in topographic mapping, and its high range accuracy (of the order of $0.1\,\mathrm{m}$) has made it suitable for civil engineering applications. It has also proved particularly useful in reconnaissance of sea ice, where a knowledge of the freeboard (the height of the surface above the water level) allows the extent of the submerged portion to be estimated. Fig. 7.4 illustrates typical output from a laser profilometer.

Laser profilers (altimeters) have not yet been flown in earth orbit, mainly because of the excessively high power requirements to generate a detectable return signal, although such an instrument was carried aboard the Apollo lunar orbiter. NASA has, however, designed an instrument (with a 700 W power consumption) which may be placed in orbit around the earth in the mid-1990s. One constraint on the use of laser pro-filometers from space is the physiological and political undesirability of subjecting human beings to excessive doses of laser radiation.

Laser ranging from ground stations to mirrors on satellites is used to determine the motion of the satellites, for the calculation of orbital parameters and the elements of the earth's gravitational field. Laser backscattering is also used for atmospheric sounding, but this application is beyond the scope of the present work.

7.3 Radar altimetry

The radar altimeter is similar in operation to the laser profilometer. The basic principle, that of timing a short pulse over its round trip from the instrument to the surface and back again, is the same. The laser profiler looks through one of the two main atmospheric windows, and the radar altimeter looks through the other one. Most of the differences between the kind of information obtainable from the two instruments can be ascribed to the larger angular IFOV of the radar altimeter, because of the longer wavelengths at which it operates.

Radar altimeters have been (and still are) employed from aircraft, as a navigational aid, for many years. They have also been used from both

Fig. 7.4. Typical output from a laser profiler (modified from Jepsky 1985).

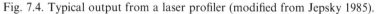
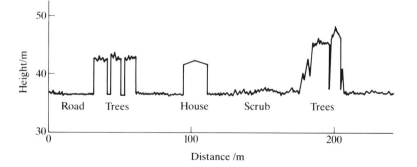

aircraft and, latterly satellites, for the investigation of land and sea surfaces.

We shall assume for the sake of simplicity that the surface being sensed is flat, and that it consists of a uniform density of isotropic, incoherent, point scatterers. (The assumption of flatness may or may not be correct, but the assumption that the scatterers are incoherent is generally false. We shall discuss the implications of this fact shortly.) We shall also neglect the operation of the inverse square law, which we can justify if the ranges of all the scatterers which make a significant contribution to the received signal do not differ significantly. As we shall see below, this is generally true.

It is apparent that the return signal, if any, received at a time t after the emission of a pulse must arise from those scatterers situated at a distance $ct/2$ from the altimeter, and it will be convenient to describe this by saying that the region from which such a return signals is received expands at a speed of $c/2$. Fig. 7.5 shows schematically that a volume whose thickness corresponds to the duration of the emitted pulse propagates away from the altimeter at this speed, and that any scatterer within this volume contributes to the received signal.

No return signal will be received until $t = t_0 = 2H/c$. A short time Δt after this, the intersection of the shaded volume of fig. 7.5 with the surface will be a disc of radius r. The power received will be proportional to the number of scatterers within this disc, i.e. to r^2. Fig. 7.6 shows that, provided that $r \ll H$, the relationship between r and Δt is

$$r^2 = cH\Delta t \tag{7.4}$$

Fig. 7.5. A radar altimeter emits a pulse of duration t_p beginning at time $t = 0$. $\Delta\theta$ is the beamwidth of the antenna. Any scatterer within the shaded region will give rise to a signal received at the antenna at time t.

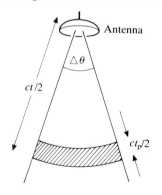

so that the returned power in fact increases linearly with time. When the trailing edge of the shaded region first touches the surface, the illumintaed area will have a radius r_{pulse} given by

$$r_{\text{pulse}} = (cHt_{\text{p}})^{\frac{1}{2}} \tag{7.5}$$

where t_{p} is the duration of the pulse. At later times, the intersection of the shaded region with the surface will be an annulus whose area, so long as its radius r continues to be very much less than the range H, has a constant value of πcHt_{p}. This simple model thus predicts that the variation of the received power with time will be as shown in fig. 7.7.

Fig. 7.6. At time Δt after the first return signal is received, the shaded region of fig. 7.5 intersects the ground in a circle of radius r. If it is assumed that ψ is much smaller than the antenna beamwidth $\Delta\theta$, and is also much less than one radian, it can be shown that $r^2 \approx cH\Delta t$.

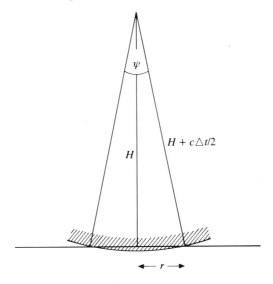

Fig. 7.7. Variation with time of the power received by a radar altimeter above a flat surface, according to the simple model derived in the text.

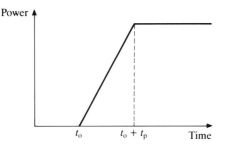

We may make our model more realistic by including the antenna power pattern, and the roughness of the surface. The power pattern of the antenna will reduce the received amplitude at large values of Δt, since these correspond to large angles with respect to the antenna's axis. Two extreme cases can be distinguished:

    (i) the *beam-limited* altimeter, for which $H\Delta\theta \ll 2r_{pulse}$

    (ii) the *pulse-limited* altimeter, for which $H\Delta\theta \gg 2r_{pulse}$

In the former case, the footprint of the instrument is governed by the half power beam width $\Delta\theta$ (in turn probably set by diffraction), whereas in the latter case we can say that most of the information about the surface is derived from the circle of radius r_{pulse}, so that the effective footprint diameter is $2r_{pulse}$. Similarly, the height resolution is $ct_p/2$.

In the case of a pulse-limited altimeter, the variation of received power with time will be similar to fig. 7.7 in the case of a flat surface. If the surface is rough, this variation will be convolved with the height distribution of the scatterers, converted to a time distribution using the factor $c/2$. The variation of the received power with time will now depend on the statistics of the surface height distribution, but it is clear that the reflected pulse will be of longer duration than the transmitted pulse. Fig. 7.8 illustrates the typical shape of a received pulse.

If t_p' is the width of the received pulse and Δh is a suitable measure of the variation in the surface height, the convolution will yield a result similar to (7.6):

$$t_p'^2 = t_p^2 + A(\Delta h/c)^2 \tag{7.6}$$

The constant A depends on the definition of Δh. For example, if the surface is that of an ocean, and Δh is the *significant wave height* (SWH) $h_{1/3}$, defined as the height exceeded by the waves one third of the time, the

Fig. 7.8. Variation with time of the power received by a radar altimeter above a rough surface, according to the model described in the text.

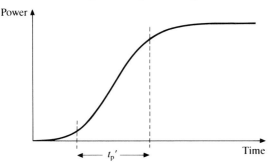

constant A can be shown to have the value of $16 \ln 2$ (≈ 11.1). Note that the increase in pulse width caused by a rough surface increases the size of the pulse-limited footprint, in accordance with (7.5). The foregoing analysis omits any consideration of the spatial roughness properties of the surface. A more sophisticated treatment is given by Brown (1977).

Let us apply this theoretical development to the radar altimeter which will be carried aboard the ERS-1 satellite, due to be launched in 1991. This instrument will operate at a frequency v of 13.7 GHz, with a bandwidth Δv of 330 MHz and a pulse width $t_p = 1/\Delta v = 3.03$ ns. The antenna diameter D will be 1.2 m and the satellite will orbit at a nominal height $H = 800$ km. Using these figures, we see that the instrument will be pulse limited, with an effective footprint diameter of 1.7 km over a smooth surface. Over an ocean surface with a SWH of 5 km, the return pulse width will become 55.6 ns, and the footprint diameter 7.3 km.

Finally, we should note that the foregoing analysis omits the effect of the earth's curvature. While this is acceptable for airborne altimeters, it is clearly a source of significant error in the case of satellite-borne instruments. However, a simple modification of (7.4) and (7.5) provides an adequate correction for this error.

If we assume the earth to be a sphere of radius R and modify fig. 7.6 as shown in fig. 7.9, we obtain by trigonometry the following expression:

$$(H + c\Delta t/2)^2 = (H + R)^2 + R^2 - 2R(H + R)\cos\alpha$$

Putting $\alpha \ll 1$, this can be simplified to give

$$\alpha^2 \approx c\Delta t/R^2[1/H + 1/R]^{-1}$$

Fig. 7.9. Modification of fig. 7.6 to allow for the earth's curvature.

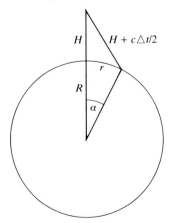

whence

$$r^2 \approx c\Delta t[1/H + 1/R]^{-1}$$

Thus (7.4) and (7.5) may still be used, as long as the altimeter's height H is replaced by an effective height $[1/H + 1/R]^{-1}$. The calculation of the time $t_0 = 2H/c$ at which the first return signal is received is, of course, unaffected by the earth's curvature.

7.3.1 Fading

Even though we have considered the effect of the antenna power pattern and have included the effect of surface roughness, our model of the operation of a radar altimeter is still crude. Clearly we can improve on the assumption that the surface consists of isotropic scatterers by incorporating the shape of the BRDF, if it is known, but there is a more important respect in which the model fails. We have assumed that the power received by the altimeter is proportional to the number of scatterers visible to it, and have thus added together the powers scattered by the various points. If the radiation reaching the radar from two points is *coherent*, that is, if there is a definite and predictable phase relationship between the two waves, the signals are capable of *interfering* with one another. They should thus be added as vector (or more precisely, *phasor*) quantities, with due regard to amplitude and phase.

Two points will be coherently illuminated with respect to each other if their distance from the source of illumination (the radar) is less than half the *coherence length* l_{coh}, and their separation perpendicular to the direction of the radiation is less than the *coherence width* w_{coh}. These quantities are given by

$$l_{coh} \approx c/\Delta v \qquad\qquad (7.7)$$

and

$$w_{coh} \approx cH/Dv \qquad\qquad (7.8)$$

where v is the frequency of the radiation, Δv is its bandwidth (which must be at least $1/t_p$, and is usually equal to it), and D is the diameter of the antenna. H, as before, is the distance from the antenna to the surface. A typical satellite-borne radar altimeter will have $l_{coh} \approx 1$ m and $w_{coh} \approx 20$ km, so that such effects are often significant. A single received pulse will, as a result of this interference, look more like fig. 7.10 than 7.7, and the phenomenon is known as *fading*. If sufficient pulses resembling fig. 7.10 are averaged, something resembling fig. 7.7 can be retrieved, and calculations of the return pulse delay time and width can be made.

7.3.2 *Applications of radar altimetry*

We have seen how the signal from a radar altimeter can be used to deduce the range H to a surface, and also to deduce the height distribution of the surface. The range H yields the surface elevation if the position of the platform is known. Since the intrinsic range accuracy of a radar altimeter can be better than 1 m, it is necessary to know the platform position to this accuracy in order to make the best possible use of the data. This is discussed in chapter 9. It is also, of course, necessary to know the speed at which radar pulses travel through the troposphere and ionosphere. If the speed of electromagnetic radiation *in vacuo* is assumed, the range will be in error by 2–3 m at a frequency of 13.7 GHz. This figure is composed of a contribution of about 2.4 m from the dry troposphere, and about 0.2 m from the ionosphere. There is also a small contribution from atmospheric water vapour. The tropospheric contribution can be modelled with sufficient accuracy, but the ionospheric contribution depends unpredictably on the time of day, year and sunspot cycle, as well as on the geomagnetic latitude. It can however be calculated from ground-based observations of ionospheric density, or multifrequency altimeter observations (which rely on the fact that the extra phase path introduced by a plasma is inversely proportional to the square of the frequency, as shown in chapter 2).

7.3.2.1 *Topographic mapping*

Topographic measurements using radar altimeters have been used to determine the shape of the *geoid* over oceans. The geoid is the surface

Fig. 7.10. Typical variation (schematic) with time of the power received by a radar altimeter above a real flat surface (compare fig. 7.7).

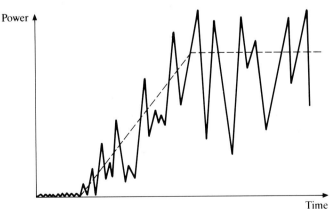

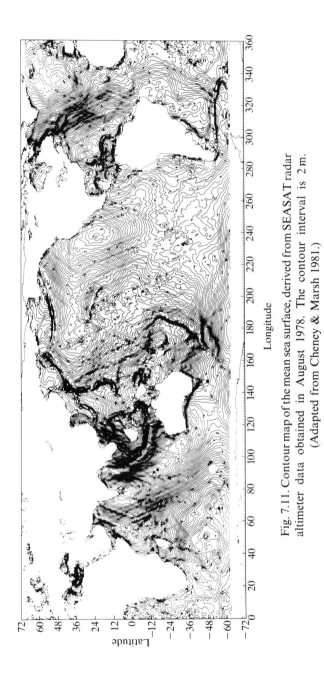

Fig. 7.11. Contour map of the mean sea surface, derived from SEASAT radar altimeter data obtained in August 1978. The contour interval is 2 m. (Adapted from Cheney & Marsh 1981.)

connecting points with the same gravitational potential, when the effects of tides, winds, ocean currents, variations in atmospheric pressure and so on have been removed. In water-covered regions, the geoid is thus in effect the average surface topography. The shape of the geoid is closely approximated by an ellipsoid of revolution, departures from this surface amounting to at most about 50 m. Tides, ocean currents and so on introduce topographic variations which are generally of the order of one metre, so that a radar altimeter with a range resolution of a few metres is capable of yielding the geoid geometry. (The converse of this is that the ultimate accuracy of a satellite altimeter is limited by knowledge of the geoid – see Wagner 1985.) Note that a small correction, arising from the asymmetry of ocean waves about their mean height, needs to be applied. This correction is termed the *sea-state bias* or *electromagnetic bias*, and is usually approximately 5 per cent of the significant wave height (e.g. Srokosz 1986).

In order to measure the quasi-static profile of the ocean surface, which is of course subject to periodic tidal variations, these variations must be averaged out unless the precision of the altimeter is not great enough to reveal them. This requires that repeated observations be made of the same region of the surface. Care must be employed in making periodic observations of a periodically varying quantity. Clearly, if the two frequencies match exactly, the varying quantity is always sampled at the same point in its cycle, so that no information is acquired about its temporal variability. This is another example of the aliasing phenomenon, which will be discussed in greater detail in chapter 9.

The average ocean surface profile, measured by the SEASAT radar altimeter (whose data, after correction, achieved a range resolution of about 0.7 m), also revealed variations in height arising from ocean currents (the surface is inclined because of the Coriolis force, an effect known as geostrophic balance), and surface manifestations of ocean-bed topography. The ocean surface above a deep submarine trench is depressed by about 15 m. (See Apel, 1983; Marsh *et al.* 1986). Fig. 7.11 shows a contour map of the mean ocean surface, derived from SEASAT data, on which these effects are clearly visible.

Topographic measurements over land surfaces are difficult to make using a satellite-borne radar altimeter. This is because the range H to the surface changes rapidly over all but the flattest terrain, and there are severe technical difficulties associated with both tracking the return pulse and recording its variation with time, so that surface roughness measurements etc. may be made. For this reason, few radar altimeter data have been obtained over land, although the terrestrial ice sheets have been studied in some detail using these instruments. SEASAT altimeter data, after correction for the satellite's orbit, were found to be consistent to ± 2.7 m on average, and to ± 0.25 m over the smoothest parts of ice sheets. The latter accuracy is approaching the level at which solid-earth tides may be observed. Two cautions must be observed in using radar altimeter data to derive the profile of an ice sheet. The first is that the instrument measures the shortest distance to the surface responsible for the echo, rather than the distance from the nadir. (At least, this is true as long as the surface slope does not exceed half the beam width of the antenna. Otherwise, the return signal may well be undetectably weak.) This effect must be corrected for, by using adjacent observations to derive the local surface slope, if an accurate topographic map is to be constructed. The second caution to be observed in using radar altimeter data for the topographic mapping of ice sheets is that, if the surface snow or ice is dry, it may be significantly penetrated by the radar altimeter signal. This may therefore lead to an overestimate of the range to the surface.

7.3.2.2 Surface roughness measurements

We have already discussed how the SWH (significant wave height) can be determined by a radar altimeter. This information is directly useful, but it can also be used to estimate the wind speed, if the fetch (the distance over which the wind acts on the water surface) is known. A combination of radar altimetry and passive microwave radiometry is capable of determining not only the SWH, but also whether the sea is fully developed, i.e. whether an increase in fetch would increase the SWH (Baskasov *et al.* 1984). Surface roughness measurements have also been made over desert areas (see Guzkowska *et al.* 1988), where the SWH is replaced by a similarly defined 'significant dune height', and may also have application to terrestrial ice sheets.

7.3.2.3 Total power measurements

The total power contained in the received pulse contains information about the BRDF of the surface, in the direction normal to the surface. This fact has been employed in interpreting radar altimeter data. Sea ice is very much smoother than an ocean surface, giving a stronger echo, and this can be used to delineate the margin of the ice pack. Wind speed can be deduced directly over the open ocean, through its influence on the surface roughness. Higher wind speeds increase the surface roughness, which in turn reduces the intensity of the return pulse. The method must be calibrated empirically, but it appears to be accurate to about $1\,\mathrm{ms^{-1}}$.

7.4 Other ranging systems

We have described the two main ranging systems which conform to the definition of remote sensing enunciated in chapter 1. We should also, however, briefly mention *radio echo-sounding*. This is a technique for measuring the thickness of ice sheets and glaciers, relying on the large attenuation length of VHF ($\approx 100\,\mathrm{MHz}$) radio waves in ice. Since the attenuation length at VHF frequencies is of the order of 100 to 1000 m (see chapter 3), it is feasible to transmit a strong signal through a body of ice, and to detect the echo from the bedrock beneath it even at a range of several kilometres which is typical of the Antarctic and Greenland ice sheets. This technique has in fact been extensively and very successfully employed (see e.g. Cracknell, 1981; Drewry, 1983; Rees, 1988) for mapping ice sheets and glaciers, with a range resolution approaching 1 metre. However, because of the long wavelength ($\approx 3\,\mathrm{m}$ in air), narrow-beamwidth antennas are not yet technologically feasible, and such remote sensing has so far been confined to observations from platforms on or relatively close to the ice surface. Satellite observations will be precluded until narrow-beamwidth instruments can be devised and placed in orbit.

Similar techniques are used for determining the thickness of saline ice, although in this case the high electrical conductivity and inhomogeneous structure of the medium greatly reduces the attenuation length. For this reason, high-power systems must be employed, and the distance from the platform to the surface must be kept small (Rees, 1988). Satellite-based remote sensing of sea-ice thickness by this technique is an even more distant prospect than for ice sheets and glaciers, whose ice is comparatively pure and homogeneous.

Finally, we should mention the use of *soil-sounding radars* in archaeological investigations. Again the technique is similar to radio echo-sounding, although the frequencies used are typical radar (microwave) frequencies rather than VHF. It has achieved a limited degree of success over dry soils in which buried artefacts produce a strong electromagnetic contrast.

Problems

1. Show that the range accuracy of an airborne laser profilometer is proportional to $H^{7/2}v^{1/2}r^{-1}$, where H is the altitude, v the velocity and r the reflectivity of the surface. Assume that the PRF is fixed and that the instrument is operated so as to obtain the maximum spatial resolution.

2. Explain the difference between *beam-limited* and *pulse-limited* radar altimeters. Show that the effective resolution of a beam-limited altimeter is $2(cH\Delta t)^{\frac{1}{2}}$, where H is the altimeter's height above the earth's surface and Δt is the duration of the received pulse. Sketch the variation with time of the power received by a radar altimeter over (a) a flat calm, (b) a rough sea.

8

Scattering techniques

8.1 Introduction

In this chapter we shall complete our description of the principal types of remote sensing instrument by discussing active systems which make direct use of the backscattered power. Optical (lidar) systems are used for sounding clouds, for discriminating between ice and water in clouds, for measuring surface albedos, aerosol and ozone profiles, for Doppler sounding of wind speed, and so on. Since most of these applications fall outside the scope of this book defined in chapter 1, however, we shall confine our attention in the present chapter entirely to radar systems. The reader interested in studying optical scattering systems should consult Chen (1985).

In section 8.2 the ground-work established in chapter 3 will be extended to a derivation of the radar equation which shows how the power detected by a radar system is related to the usual measure of backscattering ability, the differential backscattering cross-section σ^0. The remainder of the chapter describes the main types of system (and, briefly, their applications) which employ this relationship. The first and simplest is the *scatterometer*, which measures σ^0, usually only for a single region of the surface but often for a range of incidence angles. As described here, this is not an imaging system, although the dividing line between scatterometers and imaging radars is clearly a fine one.

The last two sections describe imaging radars. Section 8.4 describes the *side-looking airborne radar* (real aperture imaging radar) which achieves a usefully high spatial resolution in one dimension by time-resolving the return signal from a very short pulse. Resolution in the perpendicular

147

direction is achieved by the normal method of employing as large an antenna as possible. This approach is not feasible for satellite-borne radars, since the size of the antenna which would be required to achieve a useful resolution would be impracticably large. Instead, a large aperture (antenna) is synthesised, and the technique is thus known as *synthetic aperture radar*.

The application of imaging radar to remote sensing is still, to some extent, experimental. There is a wide range of problems to which its application is still at an early stage, and the data-processing problems inherent in the synthetic aperture approach have retarded its development. The subject is nevertheless a vast one, and an extremely comprehensive treatment of it, as well as of passive microwave remote sensing, is given by Ulaby *et al.* (1981, 1982, 1986).

8.2 The radar equation

We have already developed, in chapters 3 and 6, most of the theory necessary to calculate the response of a radar. From the definitions (3.1) of the BRDF and (3.2) of the bistatic scattering coefficient γ, we know that the power ΔP_S scattered into solid angle $d\Omega_S$, when a surface is illuminated with power P_0, is

$$\Delta P_S = P_0 \gamma \, d\Omega_S / 4\pi \tag{8.1}$$

(see fig. 8.1). γ will, in general, be a function of both the incidence and scattering directions. It is often replaced by the *backscattering cross-section* σ, defined by

$$\sigma = A\gamma \cos \theta_0 \tag{8.2}$$

where A is the area of the scattering surface and θ_0 is the incidence angle (measured from the vertical). σ thus has the dimensions of an area, and is

Fig. 8.1. Geometry for deriving the bistatic radar equation.

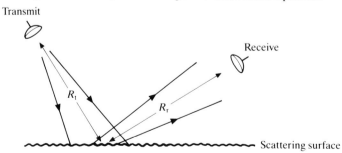

essentially the same quantity which was introduced in section 3.4.2.1. The only difference is that in the present case, σ is defined as a function of the directions of the incident and scattered radiation.

If the transmitting antenna has gain G_t and transmits power P_t, the power per unit area, perpendicular to the illumination direction, incident at the surface is (see chapter 6)

$$\frac{P_t G_t}{4\pi R_t^2}$$

The power reradiated by the surface into unit solid angle is thus, by the definition of σ, equal to

$$\frac{P_t G_t}{4\pi R_t^2}\frac{\sigma}{4\pi}$$

and, if the receiving antenna has an effective area A_r, the received power will be

$$\frac{P_t G_t}{4\pi R_t^2}\frac{A_r}{4\pi R_r^2}\sigma \tag{8.3}$$

This is the general *bistatic radar equation*. If, as is usually the case, the transmit and receive antennas are identical and (implicitly) co-located, we may simplify this to

$$P_r = \frac{P_t A_r G_t}{16\pi^2 R^4}\sigma$$

We have seen in chapter 6 that the effective area and the gain of an antenna are related by

$$G = \frac{4\pi A_e \eta}{\lambda^2}$$

where η is the antenna efficiency $(= A_e'/A_e)$ and λ is the wavelength, so we may write

$$P_r = \frac{P_t \lambda^2 G^2 \sigma}{(4\pi)^3 R^4 \eta} \tag{8.4}$$

This is in fact more commonly written in differential form,

$$dP_r = \frac{P_t \lambda^2 G^2 \sigma^0}{(4\pi)^3 \eta R^4}dA \tag{8.5}$$

where σ^0 is the (dimensionless) backscattering cross-section per unit surface area, and dA is an element of the surface area. σ^0 is often expressed in dB, and (8.5) is known as the *monostatic radar equation*.

It should be noted that in all of the foregoing we have made no mention of polarisation. In general, values of σ^0 will be defined for all four of the possible polarisation states, i.e. corresponding to horizontal or vertical polarisations of both the transmitted and received radiation. Thus it is customary to define values of σ^0_{HH}, σ^0_{VV}, σ^0_{HV} and σ^0_{VH}, where σ^0_{ij} is the value of σ^0 for radiation transmitted in polarisation i and received in polarisation j. Normally, $\sigma^0_{HV} = \sigma^0_{VH}$. In order to make the argument leading up to (8.5) as general as possible, allowance should be made for the cross-polarisation term so that (for example) a bistatic radar with a horizontal 'receive' polarisation will detect a power proportional to $P_H \sigma^0_{HH} + P_V \sigma^0_{VH}$, where P_H and P_V are the components of the transmitted power with horizontal and vertical polarisations respectively.

8.3 Scatterometry

A radar scatterometer is in general a non-imaging system which provides a quantitative measure of the differential backscattering cross-section σ^0, often as a function of incidence angle θ_0. Because of the dependence of σ^0 on surface properties, scatterometry is an especially useful technique for deducing surface roughness. It has found widespread application in characterising ocean, rock, vegetation and sea ice surfaces.

A scatterometer transmits a continuous signal or a series of pulses. The return signal is recorded, and its strength is used in conjunction with (8.3) or (8.5) to calculate the value of σ^0 for that part of the surface which is illuminated. It is particularly useful if the scatterometer can be operated in such a way as to yield the value of σ^0 as a function of the incidence angle θ_0, since this function often allows a surface material to be identified or its physical properties to be deduced. There are two principal ways of doing this. One is to use a narrow-beam scatterometer which can be steered to point at the target area. As the platform position is moved, the radar tracks the target area and the backscattering diagram is built up. The second method is to use *Doppler processing* of the signal.

Consider a scatterometer with a power pattern which is broad in the along-track direction but narrow in the perpendicular, across-track, direction. The scatterometer beam is inclined so that it looks forward. At any instant, the return signal is derived from a large range of angles $\Delta\theta$ (the beam width of the antenna), and hence from a long strip of the surface being sensed. The signal returned from the point X will be Doppler-shifted to a frequency $v_0 + \delta v$, where v_0 is the transmitted frequency and δv is given, by (2.27), as

$$\delta v = (2v_0 v/c) \sin \theta_0 \qquad (8.6)$$

This Doppler shift is unique to the incidence angle θ_0, so by feeding the return signal into a bank of filters tuned to select different Doppler shifts, data from a range of incidence angles can be extracted (fig. 8.2).

8.3.1 Applications of scatterometry

The useful output from a scatterometer, however it is realised, may be regarded as a plot of σ^0 as a function of the incidence angle θ_0 or, at least, one or more points representing this function. Since σ^0 is a modified form of the BRDF, the arguments developed in chapter 3 can be used to interpret the data. A specularly smooth surface should, in principle, show a delta function in a plot of $\sigma^0(\theta_0)$, centred at the value of θ_0 which gives specular reflexion into the radar. If the surface is horizontal, this will be $\theta_0 = 0$. A real surface is unlikely to be specularly smooth, however, particularly at the shorter radar wavelengths, and in addition the plot of $\sigma^0(\theta_0)$ will be convolved with the antenna power pattern. Thus a true delta function will never be realised, and the graph of $\sigma^0(\theta_0)$ for a smooth surface will look more like fig. 8.3.

At the opposite extreme, a Lambertian ('perfectly rough') surface has γ proportional to $\cos \theta_0$ (see chapter 3), so that σ^0 will be proportional to $\cos^2 \theta_0$. This has a characteristic shape if σ^0 is plotted logarithmically (that is, in dB). Real materials lie somewhere between these extremes. Volume scattering may also be important (we have seen how to assess this in chapter 3), in which case the variation of σ^0 with θ_0 may be even more complicated.

Even if a satisfactory theoretical model of the backscattering is not

Fig. 8.2. Doppler scatterometer. The radar emits a broad beam of width $\Delta\theta$, but radiation scattered from the point X (at incidence angle θ_0) can be identified by its Doppler shift.

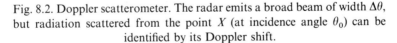

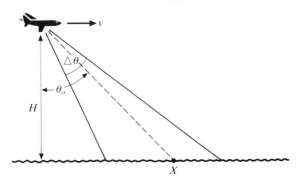

available, a plot of $\sigma^0(\theta_0)$ is often diagnostic of a surface material. Fig. 8.4
illustrates such plots for various materials.

Naturally, even more information can be obtained if observations are
made at more than one frequency or polarisation. Multiple-frequency
observations are difficult to make even from aircraft, because of the
technical complexity of providing different front ends for the radar, but
multiple polarisations (HH, VV and cross-polarised) are easier to observe
since little change needs to be made to the radar hardware.

8.3.1.1 Scatterometry over ocean surfaces

The major application of scatterometry to ocean surfaces is in deducing
wind speed and direction (although considerable work has also been
performed on the characterisation of sea ice type). This is somewhat
similar in principle to the deduction of wind speed from significant wave
height using radar altimetry, discussed in chapter 7, although more
information is available. The method relies on a model which relates the
roughness of the sea surface to the wind speed. The roughness is
anisotropic (see fig. 8.4), as would be expected since crests and troughs
tend to be perpendicular to the wind direction, and this is the key to
determining wind direction as well as speed. In fact, it has been found that
the backscattering coefficient is adequately described by the expression

$$\sigma^0 = aU^\gamma(1 + b\cos\psi + c\cos 2\psi)$$

where a, b, c and γ are constants for a given incidence angle (and
frequency), U is the wind speed, and ψ is the angle between the wind
vector and the horizontal component of the scatterometer look direction
(see e.g. Thomas & Minnett, 1986).

Fig. 8.3. Variation of σ^0 with incidence angle for a smooth surface
(schematic).

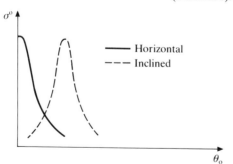

Because the statistical properties of the ocean surface are symmetrical about the wind direction, two or three observations are required, at different values of the look azimuth, to determine the wind direction. Two observations would leave an upwind/downwind ambiguity, were it not for the fact that σ^0 differs slightly in the upwind and downwind directions,

Fig. 8.4. Typical values of the dimensionless backscattering coefficient σ^0, as a function of the incidence angle θ_0, for various materials. The data are for like-polarisation, although both HH and VV data have been included. Note that most materials exhibit substantial variation about these typical values. (a) X-band; (b) Ku-band.

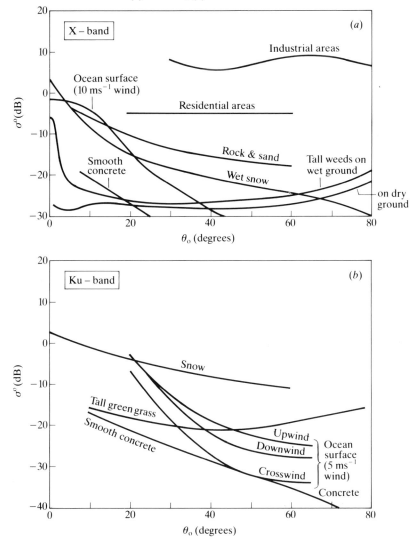

as is implicit in the above expression if $b \neq 0$. Uncertainty in the data can, however, restore the ambiguity, which may then be re-removed (or at least reduced) by a third observation.

Wind speeds were determined to an accuracy of 2 ms^{-1}, and directions to 20°, using scatterometer data from the SEASAT satellite. The same accuracy is expected to be obtained from ERS-1 data, and although the ultimate accuracy and reliability of this technique are still uncertain it is currently the most accurate way of measuring wind velocities from space. We should note, however, that the method fails to work properly during periods of rainfall. This is because of scattering from the rain itself, or anomalous scattering from a raindrop-roughened sea surface.

8.3.1.2 Scatterometry over land surfaces

Microwave scatterometry has been used extensively for characterising geological materials, using the variation of σ^0 with incidence angle θ_0 as a signature in much the same way as materials are identified in the visible band by their spectral signature. The technique has also been used for studying soil surfaces, the main retrievable parameters being moisture content, roughness and texture. It should be noted, however, that the greater complexity (compared with oceans and sea ice) of most land surfaces means that an imaging radar system is much preferable to a simple scatterometer in interpreting surface type.

Scatterometry is also beginning to find applications to the remote sensing of vegetation, particularly crops and forests. These present a substantial theoretical problem by virtue of their complicated geometry and comparatively open structure, with significant volume and surface scattering as well as, in some cases, scattering from the ground. The possibility of monitoring the extent and health of a species using this technique is an interesting one which is currently receiving much investigation, but the problem is complicated by the similarity of values of σ^0 for vegetation and for other materials. Thus high-quality scatterometers are needed if progress is to be made (see e.g. Cihlar *et al.* 1986).

8.4 Side-looking airborne radar

In sections 8.4 and 8.5 we shall discuss two types of *imaging radar*. The first is the side-looking airborne radar (SLAR), and the second, the synthetic aperture radar (SAR), is a high-resolution refinement of the SLAR.

SLAR is a *real aperture radar* (RAR) technique, so called to distinguish it from SAR methods, which is often used for scatterometry as well as for imaging. It evolved in the 1950s, as a tool for military reconnaissance, from the plan position indicator (PPI) radars developed during the Second World War. The SLAR looks to one side of the flight direction and is capable of producing a continuous strip-map of the imaged surface. It transmits short pulses of radio frequency energy, rather than a continuous wave.

Fig. 8.5 illustrates the geometry of the SLAR. The antenna is usually long and thin, mounted with its long axis parallel to the direction of motion of the platform. It looks to one side only (hence the name), so as to remove the side-to-side ambiguity in relating pulse delay to target position which would otherwise occur.

It is clear that resolution in the direction parallel to the platform's direction of motion (the *azimuth* or *along-track* direction) is achieved by virtue of the length L of the antenna. If we may write (approximately) λ/L for the angular width of the power pattern, the azimuth resolution will be

$$R_a = H\lambda/(L\cos\theta) \tag{8.7}$$

where θ is the incidence angle. This is illustrated in fig. 8.5.

Fig. 8.5. Viewing geometry of a SLAR or SAR system. The antenna is of length L and width w and is situated at a height H above the surface. It views to one side only. R_a is the resolution in the azimuth (along-track) direction at incidence angle θ_0.

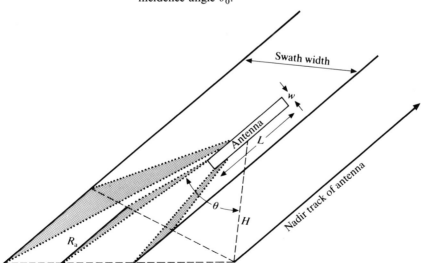

The resolution in the perpendicular (*range*, or *cross-track*) direction is determined by the pulse length τ of the radar. (Note that we are assuming the radar to be pulsed, but that this need not be so as long as some time-dependent structure is imposed on the carrier signal. The argument presented here will not be changed if we let τ stand in place of $1/\Delta v$, where Δv is the bandwidth of the transmitted signal.) The condition that two points be resolved is that their distances from the radar differ by at least $c\tau/2$, so that the range resolution R_r is given by

$$R_r = c\tau/(2 \sin \theta) \tag{8.8}$$

(see fig. 8.6).

As an example, let us consider a SLAR system operating at $\lambda = 1$ cm, with an antenna length of 5 m and a pulse length τ of 30 ns, from an aircraft at an altitude of 6000 m. 10 km from the ground track $\theta = 59°$ so $R_a = 23$ m and $R_r = 5.2$ m, whereas 25 km from the ground track $\theta = 77°$, so $R_a = 42$ m and $R_r = 4.7$ m.

We note from (8.8) that the range resolution R_r, provided that the incidence angle θ is not too small, can easily be made as small as a few tens of metres or less, and that this is independent of the platform altitude H. The azimuth resolution R_a, on the other hand, is proportional to H (8.7), and so, although a SLAR system can achieve a useful resolution for imaging from an airborne platform, it is likely to be inadequate when used from a satellite. This is the reason for the development of SAR systems, discussed in section 8.5.

8.4.1 Distortions of the image

A SLAR system measures the range to the various scattering objects within its instantaneous footprint. In the simplest form of processing, the image is presented in such a way that this range (the *slant range*) increases

Fig. 8.6. The resolution R_r in the range (across-track) direction is such that the path difference Δs is $c\tau/2$, where τ is the pulse length.

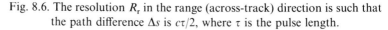
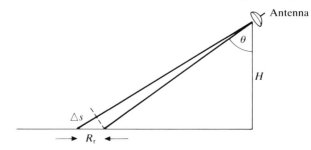

uniformly across the image. This is a form of distortion, since we actually require that the *ground range* should increase uniformly, but it can, in principle, be corrected for. Some SLAR systems incorporate a correction within the radar's signal processing unit itself.

Two further problems are caused by the topography of the surface being sensed. These are *layover* and *shadowing*, and they are consequences of the oblique angle of incidence.

Layover arises because the top of a vertical object is closer to the radar than is the bottom of the object. It is the same phenomenon as relief displacement, which was discussed in chapter 4, and for a single image it can only be corrected if the surface topography is known. It is clear that layover is more significant at small values of the incidence angle θ.

Shadowing, on the other hand, is more of a problem at large values of θ. In this case, one part of the surface is hidden from the view of the radar by another. No signal will be returned, and the corresponding part of the image will be dark. Fig. 8.7 illustrates these various types of image distortion, which can cause difficulties especially in the interpretation of geological formations from radar images (see Trevett 1986).

Fig. 8.7. Geometric distortions in SLAR and SAR images. (*a*) Slant-range distortion. Although the points *ABCDE* are equally spaced on the ground, their slant ranges *s* do not increase uniformly. Thus if the image is built up by using the pulse delay time to represent the across-track range, distortion will arise. This type of distortion can easily be corrected by calculating the corresponding ground range. (*b*) Layover. The slope *AB* appears steepened and the slope *BC* flattened. The general effect is that the hill *ABC* appears to lean towards the radar. (*c*) Shadowing. The point *C* is not visible to the radar, being obscured by the point *B*, so that it does not appear in the image. Note also (in this case) that because of the steepness of the slope *AB* the corresponding points have been imaged in the wrong order, and very close together. The latter feature means that this slope will appear particularly bright ('highlighted') on the resulting image.

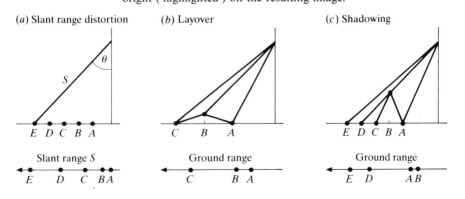

(*a*) Slant range distortion   (*b*) Layover   (*c*) Shadowing

8.4.2 *Image speckle*

In chapter 7 we mentioned the fact that coherent systems are subject to problems of fading and speckle. Here we shall treat the problem in greater detail, since the appearance of speckle on a SLAR or SAR image is particularly noticeable, and it imposes additional uncertainty on the measurement of σ^0 from the backscattered intensity.

In the remainder of this section we shall describe a simple model for the origin of speckle in coherent imaging systems. A more rigorous treatment will be found in Dainty & Newman (1986). Let us assume, as we did for the radar altimeter in chapter 7, that the SLAR system observes a flat, uniform surface consisting of isotropic point scatterers. These scatterers we shall assume to be at various heights z above some datum, in order to model a rough surface. We wish to calculate the intensity of the signal scattered from this surface in a given direction. In order to simplify the calculation we shall restrict ourselves to one dimension only.

Fig. 8.8 illustrates the geometry of the situation. The SLAR is located in the direction θ. The ray from the radar to the point P (on the surface, with coordinates x,z) and back again travels less far than the ray to and from the datum point O by twice the distance OA, which is found to be

$$OA = x \sin \theta + z(x) \cos \theta$$

Thus the phase $\phi(x)$ of the ray returning to the radar from the point x, relative to that from O, is

$$\phi(x) = -2kx \sin \theta - 2kz(x) \cos \theta$$

The amplitude received from the whole surface will be given by

$$a(\theta) = \int e^{i\phi(x)}\, dx$$

Fig. 8.8. Geometry for calculating the effect of image speckle.

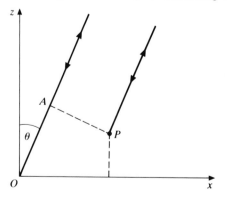

To simplify this further, we shall find the speckle pattern near $\theta \approx 0$, so that we may put $\sin\theta \approx \theta$ and $\cos\theta \approx 1$. Thus

$$a(\theta) = \int e^{-2ik[x\theta + z(x)]}\, dx$$

which is the Fourier Transform of $e^{-2ikz(x)}$. The exact nature of the speckle pattern will depend on the properties of the function $z(x)$ (in two dimensions, $z(x,y)$), which will in general be defined only statistically. Even if the r.m.s. value of $2kz(x) \ll 1$ (a very smooth surface), any plausible real function $z(x)$ will generate a distribution $a(\theta)$ which changes sign on a small angular scale. Thus the 'expected' image will be multiplied by a spatial intensity variation whose statistical properties will depend on the nature of $z(x)$. This is the characteristic speckle apparent in all coherent images (fig. 8.9), and it means that the radar power returned by an individual resolution element is subject to statistical uncertainty beyond that caused by receiver noise. This is usually undesirable, and radar speckle is often reduced by incoherent (intensity) averaging of the returns from a given resolution element. The number of independent samples averaged in this way is termed the number of 'looks'. The statistical variance of the image intensity is reduced in inverse proportion to the number of looks.

8.5 Synthetic aperture radar

The synthetic aperture radar (SAR) technique overcomes the problem set by the altitude-dependence (8.7) of the azimuth resolution of the SLAR technique. In external appearance, a SAR system is indistinguishable from a SLAR. The same geometry applies, so that fig. 8.5 describes both systems, and the same technique of emitting a short pulse and analysing the return signal by delay time is used to obtain resolution in the range direction. Higher resolution in the azimuth (along-track) direction is achieved by greater sophistication in the processing of the return signal.

Unlike the SLAR, the SAR implicitly relies on the motion of the platform to achieve high resolution in the azimuth direction. If the signal returned to the antenna over an interval of time T is stored, as amplitudes and phases, it must be possible in principle to reconstruct the signal which would have been obtained by an antenna of length vT, v being the platform speed. Since T can be made large, this 'synthetic aperture' can be made large also, thus achieving a high resolution.

Another way of thinking about this is to realise that the technique is essentially the same as the Doppler processing described in section 8.3. A given along-track coordinate on the surface will have a unique time-variation of Doppler frequency associated with it. As long as the amplitude and phase of the return signal have been recorded, this component may be extracted. The process can then be repeated for another value of the along-track coordinate.

We thus see that a SAR will have a better azimuth resolution than a SLAR. We can estimate the best possible resolution as follows:

Fig. 8.9. An L-band SAR image showing the phenomenon of 'speckle' as a characteristic graininess. The image was obtained in 1978 by the SEASAT satellite from an altitude of approximately 800 km, and shows an area of about 37 × 40 km near Dunkirk, France. Note the features visible in the water, which are probably a manifestation on the surface of the bottom topography. Note also the many features of the land surface which are visible, including roads, rivers and canals, and built-up areas. The latter are particularly clearly imaged, showing as concentrations of high backscattering. (Reproduced by courtesy of the National Remote Sensing Centre, UK.)

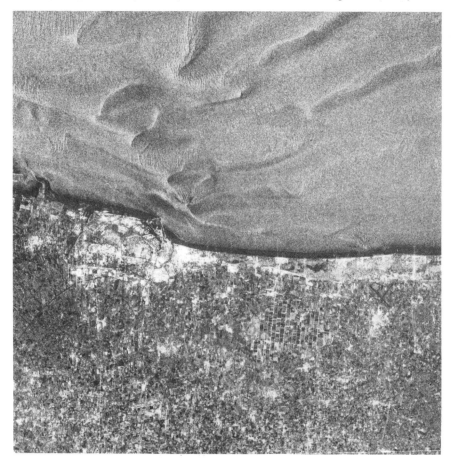

Let us assume for simplicity that the SAR is imaging a strip vertically below, at a distance H. If the length of the (real) SAR antenna is L, the beamwidth will be approximately λ/L, which corresponds to a distance $\lambda H/L$ along the track. A given point will thus be illuminated by the antenna only during the interval in which the radar travels this distance, so this is the maximum useful synthetic aperture length. The angular resolution of this synthetic aperture will be approximately L/H, giving a surface resolution of L.

This calculation is only approximate, but it illustrates two important points. The best achievable resolution is proportional to L. (In fact, it is $L/2$ as we shall see in section 8.5.1. Note that most derivations which follow the lines of the preceding argument fudge this factor of 2.) It is independent of the platform height H. Both of these are contrary to our experience of other imaging systems. The latter point shows that our assumption (which we introduced for simplicity) that the point being imaged is vertically below the radar, is irrelevant, and that the best achievable resolution will be the same at all values of the incidence angle.

8.5.1 More exact treatment of a SAR system

In this section we shall derive somewhat more rigorously the result that the best possible azimuth resolution of a SAR is $L/2$. We shall again make the temporary simplifying assumption that the point being observed is directly below the radar; as before, this will turn out not to matter since the resolution is independent of the range to the target.

Fig. 8.10 illustrates the necessary geometry. The antenna has an aperture distribution $a(y)$, referred to its centre at $y=0$. The coordinate z measures the instantaneous position of the antenna centre A, and we shall assume that the data recorded by the antenna will be processed in such a way as to *focus* on the point $x=0$. This is clearly done by altering the phase of the signal recorded at position z to take account of the path length OA.

The signal $A(x,z)$ received from a point with ground coordinate x, when the antenna is at position z, will then be proportional to

$$A(x,z) = F(\beta)\exp(2ik\Delta r) \qquad (8.9)$$

where $F(\beta)$ is the amplitude diffraction pattern of the antenna. The latter is given by

$$F(\beta) = \int_{-\infty}^{\infty} a(y)\exp(iky\beta)\,dy, \qquad (8.10)$$

where we have assumed that $\sin \beta \approx \beta$, in which case we may also put

$$\beta = (z - x)/H \qquad (8.11)$$

The total amplitude $A(x)$ received from the point x is obtained by adding together the values of $A(x,z)$ obtained at all positions z of the antenna, i.e.

$$A(x) = \int_{-\infty}^{\infty} A(x,z)\, dz \qquad (8.12)$$

Substituting into (8.12) from (8.9), (8.10) and (8.11), and putting $\Delta r = x\beta = xz/H$, gives

$$A(x) = H \int_{-\infty}^{\infty} F(\beta)\exp(2ikx\beta)\, d\beta$$

$$= a(-2x) \qquad (8.13)$$

by comparison with (8.10), and ignoring constant factors outside the integral. Thus the response on the ground has exactly the same shape as the antenna aperture function, but is half as wide. We may therefore write the focussed azimuth resolution as

$$R_a\ (\text{focussed}) = L/2 \qquad (8.14)$$

The way in which the processing of SAR data is carried out is described in detail by Raney (1982). Nowadays it is normally performed digitally,

Fig. 8.10. Geometric construction for calculating the azimuth resolution of a SAR. The antenna's (shaded) position is defined by the coordinate z. The data are processed in such a way as to focus the antenna on the point $x=0$. The text shows how the response to the point P is calculated.

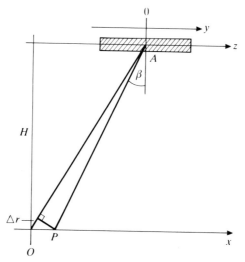

but in the early 1960s, when digital computers were very much less powerful than they are now, it was performed optically. The signals were recorded on film and, in effect, used as a diffraction mask to focus light in such a way as to reconstitute the original scene. Much of the rather complicated theory of this procedure, which space does not permit us to discuss here in any detail, was developed by E.N. Leith, who also made substantial contributions to the development of the closely related field of holography.

8.5.2 Unfocussed SAR

A much simpler form of processing ignores the variation of the length OA with position z of the antenna. This is termed *unfocussed SAR*. Clearly, in order to image a point at $x=0$, it is essential to add together signals obtained from a more limited range of values of z, namely those for which the length OA does not exceed H by more than (say) $\lambda/8$. Pythagoras' theorem shows that, for large H, the synthetic aperture will then have a length of $(\lambda H)^{\frac{1}{2}}$, so that the ground resolution is also equal to this value:

$$R_a \text{ (unfocussed)} = (\lambda H)^{\frac{1}{2}} \tag{8.15}$$

(Note that definitions of the resolution of an unfocussed SAR vary, in the sense that our choice of the $\lambda/8$ criterion is somewhat arbitrary. However, no realistic definition of R_a (unfocussed) will differ from (8.15) by a factor of more than about 2.) In this case, the resolution is independent of the antenna size, and is degraded by large values of H or λ. The latter feature is more in keeping with our experience of imaging systems, although the former is still unexpected.

8.5.3 Limitations imposed by ambiguity

The complicated manner in which the pulses received by a SAR are processed in order to achieve high resolution leads to a rather curious constraint on the size of the antenna (see Elachi 1987). Because the system is pulsed, the pulse repetition frequency (PRF) p must be low enough to avoid range ambiguities (see chapter 7). By referring to figs. 8.5 and 8.6, we can see that for this to be the case,

$$p^{-1} > (2H/c)(\sec\theta_{max} - \sec\theta_{min})$$

where θ_{max} and θ_{min} describe the limits of the antenna power pattern and c is the velocity of light. If we assume that the antenna beamwidth

$\Delta\theta = \theta_{max} - \theta_{min}$ is small, we can write

$$(\sec\theta_{max} - \sec\theta_{min}) \approx \Delta\theta \sin\theta/\cos^2\theta,$$

and $\Delta\theta \approx \lambda/w$, so that

$$p < \frac{cw}{2\lambda H}\frac{\cos^2\theta}{\sin\theta} \tag{8.16}$$

On the other hand, if the PRF is made too small, the return signal will be inadequately sampled to obtain the desired azimuth resolution. Since this is $L/2$, the PRF must satisfy

$$p > 2v/L \tag{8.17}$$

where v is the platform velocity, in order not to degrade this resolution. Combining the inequalities (8.16) and (8.17) gives

$$wL > \frac{4v\lambda H}{c}\frac{\sin\theta}{\cos^2\theta} \tag{8.18}$$

Thus we see that the swath width and azimuth resolution cannot be varied independently of one another for a given SAR system, and that a given wavelength, look angle, platform velocity and height impose a minimum value on the area of the antenna. The actual size of the antenna may of course be larger than this minimum value, in order to radiate sufficient power.

8.5.4 Imaging of moving targets

In describing the way in which a SAR achieves high resolution in the azimuth (along-track) direction, we have assumed that the target is stationary. If it is moving, the complicated way in which the data are processed results in a shift in the apparent position of the object. This can be amusing or useful, for example in determining the speed of a railway train from its apparent displacement from the tracks, but it has an adverse and imperfectly understood effect on the imaging of water waves. This is at present hampering efforts to measure directional wave energy spectra from SAR images (see Ouchi 1986).

We can understand this image displacement by considering the SAR imaging process as a Doppler frequency analysis. As the radar approaches a stationary target, the Doppler shift decreases, reaching zero when the radar has achieved the same along-track position as the target. If the target is in motion, an extra Doppler shift will be added to that due to the motion of the platform. The Doppler shift will thus fall to zero at a different value of the along-track coordinate, and the processor will assign this value to the along-track position of the target.

Fig. 8.11 shows how we can calculate this shift. We shall assume that the target is located at the origin of a Cartesian coordinate system, moving with velocity $(u \cos \psi, u \sin \psi, 0)$. The radar is located at position $\mathbf{z} = (x, y, H)$ and has velocity $(0, v, 0)$. The relative velocity $\mathbf{v}'$ of the target with respect to the radar is thus

$$\mathbf{v}' = (u \cos \psi, u \sin \psi - v, 0)$$

The Doppler shift will fall to zero when the vectors $\mathbf{v}'$ and $\mathbf{z}$ are perpendicular (2.27), i.e. when $\mathbf{v}' \cdot \mathbf{z} = 0$. Thus the y-coordinate of the radar when this happens is

$$y = \frac{ux \cos \psi}{v - u \sin \psi} \tag{8.19}$$

This is the value of the y-coordinate which will be ascribed to the target. If $v \gg u$, this simplifies to

$$y \approx (ux/v) \cos \psi \tag{8.20}$$

that is, the along-track displacement depends on the across-track component of the target's velocity, its ground range, and the platform velocity. Fig. 8.12 shows an example of azimuth shift in a SAR image.

This shift will, in general, only be apparent if the motion of the target is maintained throughout the time during which the target is illuminated by the radar. This period is called the *coherence time* of the system, and it may easily be seen, by a simple extension of the arguments developed earlier in this section, that it must be about

$$\lambda H / Lv$$

Fig. 8.11. Construction for determining the azimuth shift in a SAR image of a target moving with speed u in the direction ψ.

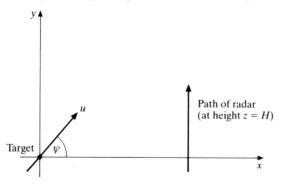

8.6 Applications of radar imaging

SLAR and SAR images are generally produced in a format similar to that of an aerial black and white photograph, the brightness of the displayed image being dependent on the value of σ^0. Such images may be visually interpreted singly or as stereo pairs. However, increasing use is now being made of digital imagery which can be analysed by the techniques of digital image processing (chapter 10), and which make the greatest possible use of the quantitative nature of the data.

Radar images (from SLAR or SAR systems) are modulated by the same processes which govern scatterometer data, and they find applic-

Fig. 8.12. Azimuth shift in a SAR image. The image of the moving ship (the bright rectangular region, left centre) is displaced from the image of its wake (the dark diagonal stripe). (Reproduced from Wahl *et al.* (1986) by courtesy of the European Space Agency.)

ation in broadly the same areas. However, the usual advantages (especially to a human interpreter) of presenting the data as a two-dimensional image apply. The image can be assessed and classified by eye, and spatial relationships, discussed in greater detail in chapter 10, are revealed. Geometrical and topographic effects such as highlights, shadows and layover, and the coherent phenomenon of speckle, may make the image more difficult to analyse, as may the fact that volume scattering may contribute significantly to the detected signal. The great complexity of processing SAR data is also a factor not in its favour, and for these various reasons the application of SAR data to many tasks is still at a comparatively early stage. It is still a lengthy job to process a SAR image, taking of the order of one hour (on a CRAY computer) to process a 100 km square image. On the other hand, the availability of calibrated backscattering data, possibly at a number of polarisations and frequencies, is of great potential value. When this is coupled with the fact that radar signals are unaffected by darkness, and largely unaffected by most forms of weather, the technique becomes very attractive. Wooding (1988) presents a useful summary of applications of imaging radar.

Because of the complexity of the techniques involved, imaging radar data are expensive to acquire. This has in the past imposed an economic constraint on the development of applications of such data, and the earliest civilian use was in the potentially profitable area of geological exploration. This has latterly grown into applications in geomorphological mapping and analysis (see Trevett 1986). Other important applications are, as discussed earlier under scatterometry, the investigation of soil and crop conditions, and of other types of vegetation. Plate 8 shows an example of a bispectral radar image of a land surface. Microwave radiation can penetrate significant distances into arid surfaces (see the discussion of soil sounding radars in chapter 7), and SAR images of the Sahara desert have revealed traces of old water courses.

Imaging radar has found wide application in oceanography. Surface wave fields are imaged distinctly, despite the difficulties, mentioned in section 8.5.4, of deducing wave spectra from SAR images. Diffraction of waves by coastal features, and refraction by variations in bottom topography (e.g. fig. 8.9), are often clearly visible. There is also some evidence for the imaging of internal waves. Small-scale roughness is reduced by the presence of natural and artificial slicks, and these have been detected in radar imagery.

Imaging radar has also been extensively applied to the study of sea ice,

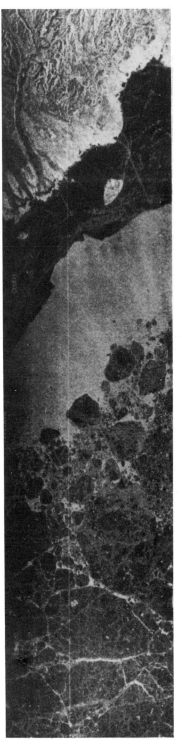

Fig. 8.13. SEASAT SAR (L-band) image of part of the Beaufort Sea west of Banks Island, Canada (right). The image was obtained on 11 July 1978, and covers an area approximately 30 by 120 km. From right to left, it shows stream channels, alluvial fans and beaches on the island, followed by a dark region which is shore-fast ice 1–2 metres thick. This ice contains a number of ridges (bright linear features). The next zone (grey) is a region of open water, followed by unconsolidated and finally consolidated pack ice. Within the pack-ice floes ridges can be distinguished, as well as bright areas of very rough ice. (Reproduced by courtesy of the National Aeronautics and Space Administration, USA.)

where its ability to penetrate darkness and cloud is particularly advantageous. The delineation of boundaries between ice floes and open water is usually straightforward (see fig. 8.13), and comparison of consecutive images of the same area of ocean allows the motion of ice floes to be tracked. A rather more difficult problem, to which a considerable amount of attention is presently being paid, is the determination of ice type from radar images. As is often the case, the problem is that the single observable quantity σ^0 depends on a large number of parameters.

It is likely that the range of applications of imaging radar data will increase greatly in the near future. A potent spur to this development is the emphasis which many national and international space agencies are now placing on microwave (active and passive) remote sensing: many SAR systems will be placed in space before the year 2000.

Problems

1. A radar antenna has an effective area of $10\,\mathrm{m}^2$, and an efficiency of 0.7 when operated at a wavelength of 10 cm. By what factor is the power received from an area 30 m square, having a value of σ^0 of 0 dB, less than that transmitted, if the range is 800 km?

2. A very simplified model of the radar backscattering coefficient of a sea surface is

$$\sigma^0 = A + B\cos 2\psi$$

where ψ is the angle between the wind direction and the radar look azimuth, and A and B are as follows for Ku band scattering at 40° incidence angle, HH polarisation:

$$A = 0.8v - 30$$
$$B = 3.5 - 0.1v$$

v is the wind speed in ms^{-1} and σ^0 is given by these expressions in dB.

   A scatterometer observation measures σ^0 values of $-22.9\,\mathrm{dB}$ and $-21.1\,\mathrm{dB}$ looking north and east respectively. Find the wind velocity. Is there any ambiguity in your answer? If so, could it be removed by a third observation?

3. Show that the maximum azimuth resolution of a SAR system is $L/2$ for a focussed system, and $\approx (\lambda H)^{\frac{1}{2}}$ for an unfocussed system, where L is the antenna length, λ is the wavelength and H is the height of the SAR above the ground. Estimate the ratio of the processing times

9

Platforms for remote sensing

9.1 Introduction

In this chapter we shall consider aircraft and satellites as platforms for remote sensing. Other, less commonly used, means of holding a sensor aloft are also available (e.g. towers, balloons and kites), but we will not discuss these. The reason for this, apart from their comparative infrequency of use, is that most remote sensing systems make direct or indirect use of the relative motion of the sensor and the target, and this is more likely to be available and controllable in the case of the types of platform to which we are restricting ourselves. Fig. 9.1 shows schematically the range of platforms and heights which is available.

The spatial scale and temporal variability of the phenomenon to be studied will determine the observing strategy to be employed, and this will have an influence on the choice of operational parameters in the case of an airborne observation, or on the orbital parameters in the case of a satellite-borne sensor. Clearly, these considerations will also place limits on the type of sensor to be employed. Fig. 9.2 illustrates this idea in a schematic way.

9.2 Aircraft

Aircraft of various types provide an exceptionally convenient platform for remote sensing. With a suitable choice of vehicle a range of heights can be covered from a few dozen metres to many kilometres, above most of the earth's atmosphere. The range of speeds available is more or less continuous from zero (in the case of a hovering helicopter) to several

171

hundred metres per second. The capacity for carrying equipment can range from 50 kg or so to many tons.

What determines the choice of operational conditions for an airborne sensor? We may group the operational aspects into two sections – intentional operational decisions, and the effects of errors and perturb-

Fig. 9.1. Remote sensing platforms, by height (schematic). (Reproduced by courtesy of the National Remote Sensing Centre, UK.)

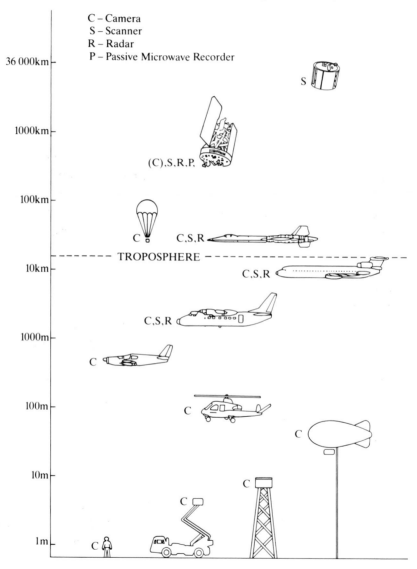

ations in the motion. Into the first category we put the choice of height (which determines the scale, coverage and linear resolution of imaging sensors, and has an analogous role for non-imaging sensors), speed (which determines the linear sampling rate) and route. This last mentioned consideration may not be trivial in the case of sensors which look obliquely, particularly the SLAR and SAR imaging radars. Clearly to acquire information from a given location, the choice of route and incidence angle will be interdependent, and there may well be, as we have seen in the case of the imaging radars, good scientific reasons for preferring particular incidence angles over others.

Into the latter category, that of unintentional variations in the motion of the aircraft, we can put errors of position, and irregularities in the motion of the aircraft. The implication of a position error is fairly obvious – the image is of somewhere other than intended. There are two approaches to this kind of error, either or both of which may be employed. The first approach is to use electronic navigation systems to allow the subsequent determination of the aircraft's true position. These systems use either surface based (e.g. LORAN, OMEGA) or satellite based (e.g. SATNAV, GPS) radio transmitters to allow the calculation of position by ranging and triangulation, and they provide accuracies

Fig. 9.2. Observational requirements (resolution and repeat period) for various disciplines. (Adapted from Fraysse, 1984.)

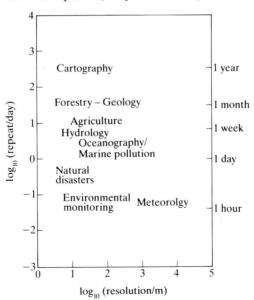

ranging from a few hundred metres to less than one metre. The second approach to the problem of position errors is to use *ground control points* (GCPs). These are features of known location which can be seen in the image and used to correct its position and orientation (and scale, if necessary). They can be 'naturally' occurring (including man-made), or emplaced especially for the purpose. If of the former kind they might be, for example, a cross-roads whose grid reference is known, or the outline of a coast. If of the latter kind they could be a suitable marker fixed to the ground or, for active microwave sensors, a radar transponder of some kind.

Irregularities in the aircraft's motion, such as roll and yaw, are often compensated for by sensors, particularly in imaging radar systems. However, if the disturbance is severe (for example during strong turbulence or vigorous manoeuvring) there may be significant residual effects. In scanned systems these will show up as distortions in the image (see Cracknell 1981). Such distortions can sometimes be removed if the perturbations are known, or if there are sufficient GCPs in the image (see chapter 10).

9.3 Satellites

Placing a satellite in orbit about the earth is clearly more expensive than mounting an airborne remote sensing campaign, but the advantages in terms of increased platform speed and swath width, as well as continuity of mission, are substantial. In general, the data coverage obtained from a satellite mission is better than that obtainable from an airborne mission, and the fact that a satellite may continue to function for five years means that the data are also homogeneous. The cost-effectiveness of satellite remote sensing was discussed briefly in chapter 1, where it was pointed out that the economic advantages of an operational satellite (rather than one launched purely for research purposes) justify the cost of launching it. An obvious advantage of using satellites for remote sensing, but one which can nevertheless pose interesting legal problems, is the fact that the laws of orbital dynamics do not respect political boundaries. Rao & Chandrashekar (1986) have discussed some of the legal implications of satellite remote sensing, and it is also discussed in Cracknell (1981). Bondi (1988) gives a particularly clear and simple example of the legal and moral problems which can be raised by the question of the ownership of knowledge obtained from spaceborne remote sensing.

9.3.1 *Launch of satellites*

In this section we give a brief indication of the considerations which apply to the placing of a satellite in orbit about the earth. A detailed treatment is not appropriate, but the interested reader is referred to Massey (1964) and to Chetty (1988).

To place a satellite in a stable (apart from the effects of air friction and other perturbations) orbit, it is necessary to overcome the earth's gravitational attraction and, to a lesser extent, the resistance of the lower atmosphere to the passage of a body. This is achieved using a *rocket*, which is a vehicle which carries all its own fuel (including the oxidising agent), deriving a forward thrust from the expulsion backwards of the combustion products.

Elementary classical mechanics shows that a rocket of total initial mass M_i burning a mass M_f of fuel will increase its velocity, in the absence of gravitational and friction forces, by

$$\Delta V = U \ln(M_i/(M_i - M_f)) \qquad (9.1)$$

where U is the velocity of the exhaust gases with respect to the rocket. Since the orbital velocity of a satellite in a low earth orbit is about $7\,\mathrm{km\,s^{-1}}$ and a typical value of U is only $2.4\,\mathrm{km\,s^{-1}}$, we can estimate that a rocket capable of reaching such an orbit must initially consist of at least 95% fuel. If the effects of gravity and air drag are included, this figure is increased to 97%. Naturally, the payload itself can represent only a small fraction of the remaining 3%, and in consequence single-stage rockets are capable of placing only very small masses into orbit. Instead, multiple-stage rockets, with three or four stages, are used, and these are capable of putting payloads of a few tons into low earth orbits, and smaller payloads into geosynchronous orbits. The space shuttle is at present capable of placing about 30 tons of payload into a 400 km orbit, and about 6 tons in a geostationary orbit.

Once the satellite has been placed in its desired orbit, a limited amount of manoeuvring may still be possible if it carries thrusters. These are small controllable rockets, giving small thrusts, and they may be used, for example, to move the satellite into an adjacent orbit, or to rectify the effects of perturbations caused by friction, gravitational irregularities, solar pressure and so on. Because of these perturbations, whether corrected or not, it is impossible to predict the position of a satellite with sufficient accuracy for most remote sensing applications (especially range measuring systems such as the radar altimeter). For this reason, the satellite's position must be continually redetermined, usually by laser or radio ranging techniques. In fact, such measurements are often used to deduce the perturbing forces acting on the satellite, and hence (e.g.) the earth's atmospheric density (see section 9.3.5, and Massey 1964) and gravitational field. Ranging to a satellite from surface stations has also been used to monitor tectonic motion (e.g. Christodoulidis *et al.*, 1985).

9.3.2 Description of the orbit

If the earth were a spherically symmetric mass with no atmosphere, the motion of a satellite of negligible mass would be in an elliptical orbit with the earth's centre as one focus. It will be useful to begin our investigation of the motion of satellite orbits by considering this idealised case, and fig. 9.3 illustrates the terms we shall need to describe it.

The equation of the ellipse is

$$r = a(1-e^2)/(1+e\cos\theta) \tag{9.2}$$

and the period of the orbit can easily be shown, using Newton's law of gravitation, to be

$$P_0 = 2\pi a^{3/2}(GM)^{-1/2} \tag{9.3}$$

where G is the universal gravitational constant and M is the earth's mass. Although neither G nor M has yet been measured particularly accurately, their product GM has been measured to high precision by observing the orbits of artificial satellites about the earth. The best recent value (Smith *et al.* 1985) for this constant is

$$GM = (3.98600434 \pm 0.00000002) \times 10^{14}\,\mathrm{m^3\,s^{-2}}$$

Fig. 9.3. Description of an elliptical orbit about the earth. S is the satellite, E the earth's centre, P the perigee of the orbit and A the apogee. a and b are the semimajor and semiminor axes, respectively, and e is the eccentricity of the orbit, given by $e = (1-b^2/a^2)^{1/2}$. The instantaneous position of the satellite is specified by the angle θ and the radius r.

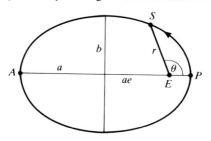

Although (9.2) describes the shape of the orbit, it does not describe the satellite's motion dynamically, that is, it does not tell us how θ varies with time t. This is in fact rather harder to do, and the result is not particularly simple (except when $e=0$, in which case it is trivial). It is

$$\frac{t}{P_0}=\frac{1}{\pi}\arctan\frac{(1-e)\tan(\theta/2)}{(1-e^2)^{\frac{1}{2}}}-\frac{e}{2\pi}\frac{(1-e^2)^{\frac{1}{2}}\sin\theta}{1+e\cos\theta} \qquad (9.4)$$

Note that this expression is the wrong way round for most purposes. It gives the time as a function of the angle, whereas we usually want to know the angle as a function of the time. Unfortunately (9.4) is not invertible, in the sense that it cannot be rewritten as an expression for θ in terms of t.

In fact, most artificial satellites are placed in orbits with very low eccentricities, usually 0.01 to 0.001, and we may continue for the time being to develop our description of orbital motion on the assumption that $e=0$. The orbit must be concentric with the earth (which we are still assuming to be spherically symmetrical) but need not be parallel to the equator. The angle between the plane of the orbit and the plane of the equator is called the *inclination* of the orbit (see fig. 9.4). The inclination i is conventionally always positive, and is less than 90° if the satellite's orbit is *prograde* (i.e. in the same sense as the earth's rotation about its axis), greater than 90° if the orbit is *retrograde*. Clearly, a truly polar orbit, in

Fig. 9.4. The orbit of an artificial satellite in relation to the earth's surface. The inclination i is the angle between the plane of the orbit and the equator. Ω is the ascending node of the orbit, where it crosses the equator in a northerly direction, and ϕ is the angular position of the satellite S measured from the centre of the earth E. NP and SP are the earth's north and south poles. The curved arrow near the north pole shows the direction in which the earth rotates.

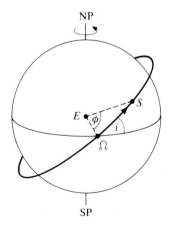

which the satellite passes directly over the poles, has $i = 90°$. Near-polar orbits give the greatest coverage of the earth's surface, and are used for low-orbit (i.e. not geostationary) meteorological satellites such as NIMBUS. It is however in general more costly to put a satellite in a near-polar orbit, because less advantage can be taken of the earth's rotation during the launch phase.

The most important descriptor of the satellite's motion after its height above the surface is the path traced out by the sub-satellite point on the surface of the earth. This can be described by calculating the latitude b and longitude l of this point using spherical trigonometry. The point Ω in fig. 9.4 is called the *ascending node* of the orbit, and is the point where the northbound satellite crosses the equator. If the angle subtended at the centre of the earth between Ω and the satellite is ϕ (fig. 9.4), and the instantaneous longitude of Ω is l_0, the position of the sub-satellite point is given by

$$\left.\begin{array}{l} \sin b = \sin \phi \sin i \\ l = l_0 + \Delta l \\ \cos \Delta l = \cos \phi / \cos b \end{array}\right\} \tag{9.5}$$

To interpret this set of equations, we need to add the condition that Δl and ϕ be in the same quadrant if $i < 90°$, and in different quadrants if $i > 90°$. This removes the ambiguity inherent in taking the inverse cosine. The convention adopted for the signs of b and l is the Cartesian one, that b is positive in the northern hemisphere and l increases to the east.

Because of the earth's rotation, the sub-satellite track will not be a great circle. When the satellite has completed one orbit, the earth will have turned to the east and so the orbit will appear to drift to the west. Note that this is true of both prograde and retrograde orbits. The rotation of the earth may be taken into account in equations (9.5) by realising that it is equivalent to a uniform rate of change of l_0. However, before we are tempted to apply this simple precept we must discuss the topic of orbital precession.

9.3.3 Precession

Thus far we have assumed the earth to be spherical. It is not; it is, roughly speaking, an oblate spheroid (i.e. the equator bulges outwards). The most convenient way to describe mathematically the effect of this non-spherical earth on the motion of a satellite is to write the gravitational potential as a (possibly infinite) sum of spherical harmonics. As we might expect, the azimuthal (longitudinal) variation is small, and it will normally be adequate to expand the expression to the second order in the latitudinal terms only:

$$V = -GM/r[1 - a_e^2 J_2/2r^2 \, (3\sin^2 b - 1) + \ldots]$$

where a_e is the earth's equatorial radius. The non-zero term J_2, which expresses the equatorial bulge and which has a value of

$$J_2 \approx 1.08263 \times 10^{-3}$$

is called the *dynamical form factor* and has three important effects:
Firstly, it causes the orbital plane to rotate (*precess*) about the polar axis. This occurs at an angular speed of

$$\Omega = -3J_2/2 \, (GM/a)^{\frac{1}{2}} (a_e^2/a^3) \cos i/(1 - e^2)^2 \qquad (9.6)$$

As before, the convention is that if $\Omega > 0$ the precession is prograde. This clearly only occurs if $i > 90°$, i.e. when the orbit itself is retrograde. This precession is illustrated in fig. 9.5.

Fig. 9.5. Effects of the earth's equatorial bulge on satellite orbits. (*a*) Precession of the orbital plane about the earth's polar axis. (*b*) Rotation of the orbit in its own plane.

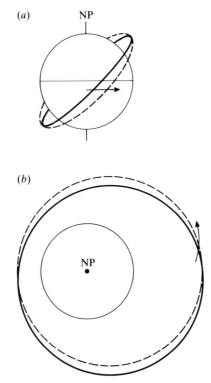

Secondly, the nodal period (i.e. the time between successive ascending or descending notes) is increased from P_0 to P_N, where

$$P_N = 2\pi a^{3/2}(GM)^{-\frac{1}{2}}\left[1+\frac{3J_2}{4}\left(\frac{a_e}{a}\right)^2\left\{(1-3\cos^2 i)+\frac{(1-5\cos^2 i)}{(1-e^2)^2}\right\}\right]$$

(9.7)

Thirdly, the ellipse (if the orbit is elliptical) itself rotates in its own plane, at a rate

$$\omega = \Omega(1-5\cos^2 i)/2\cos i$$ (9.8)

This motion is also illustrated in fig. 9.5.

9.3.4 Special orbits

The dependence of the dynamic behaviour of a satellite in orbit about the earth on comparatively many parameters creates the possibility of 'tuning' these parameters to give particularly useful orbits. In this section we will discuss some of the most important of these special orbits.

(a) Geostationary orbits

A satellite in a geostationary orbit is, as its name suggests, at rest with respect to the rotating earth. This is brought about by putting the satellite into a circular orbit above the equator, with a nodal period P_N equal to the earth's rotational period P_E. The period P_E is not equal to 86 400 seconds (24 hours) as one might at first imagine. This is because in 24 hours the earth rotates once with respect to the sun, but since it is also orbiting the sun in the same direction as it rotates, it has in fact rotated by slightly more than one complete turn with respect to the 'fixed stars'. In fact in 24 hours the earth will have rotated $1+1/365.25$ complete turns, so P_E must be $86\,400 \times 365.25/366.25$ s, or about 86 164 s. This is called a *sidereal day* (from the Latin *sidus*, a star), and is the time taken for the earth to rotate once with respect to the fixed stars. Thus for a geostationary orbit we must have $i=0$, $e=0$ and $P_N = P_E = 86\,164$ s. If we take the earth's equatorial radius as

$$a_e = 6.378135 \times 10^6 \text{ m}$$

(9.7) gives $a=42\,170$ km. Thus geostationary satellites are about 35 800 km above the equator. Such orbits are used by the METEOSAT and GOES weather satellites, giving continuous coverage of a large part of the earth's surface (see fig. 9.6), and by telephone and television relay satellites.

(b) Geosynchronous orbits

The main defect of the geostationary orbit for relay satellites is that it is located over the equator. This means that the line of sight to the satellite from a point on the earth's surface at a high latitude will have a low elevation angle, passing through more of the atmosphere than is desirable. Indeed, if the latitude is too great the satellite will be below the horizon. (The useful coverage of a geostationary satellite is usually reckoned to be a small circle of radius 55°, centred at the sub-satellite point, for quantitative analysis and of 65° for qualitative work.) What

Fig. 9.6. View of the earth from a geostationary satellite. This image was recorded by a METEOSAT satellite located at longitude 0° and above the equator, and it shows radiation from the earth in the band 0.4–1.1 μm. (Reproduced by courtesy of the National Remote Sensing Centre, UK.)

would be desirable, but of course impossible, would be to place a satellite in a geostationary orbit above an arbitrary point, not necessarily on the equator. Instead, the best that can be done is to use a geosynchronous orbit, which has $P_N = P_E$ but $i \neq 0$. The sub-satellite path traces a figure-of-eight pattern, crossing at a fixed point on the equator and reaching a maximum latitude of $\pm i$ (fig. 9.7). The semimajor axis of this kind of orbit is almost the same as that for a geostationary orbit.

A satellite in a geosynchronous orbit with $i \neq 0$ traces its figure-of-eight at an approximately uniform rate, and this means that for roughly half a day its position is actually less useful than if a geostationary ($i = 0$) orbit had been used. That is, for half a day the satellite is above (say) the Northern hemisphere, when its intended use is in the Southern hemisphere. An ingenious solution to this problem is the use of the *Molniya orbit*, (from the Russian МОЛНИЯ, lightning) which is highly eccentric with a minimum distance above the earth's surface of only about 500 km. The apogee is positioned above the desired point. The orbital period is chosen to be *half* a sidereal day. Since by Kepler's second law the satellite orbits at an angular speed inversely proportional to its radial distance, it remains 'on station' for a considerably longer time, quickly passing through the loop of the figure-of-eight which is in the wrong hemisphere (i.e. on the perigee side). We can calculate these times using (9.4). Since the perigee distance is about 6900 km, the eccentricity e must be about 0.74. Putting this value and $\theta = \pi/2$ into (9.4), we find that $t/P_0 = 0.04$. Thus the satellite spends 8% of its time in the 'wrong' hemisphere, and 92% in the 'correct' one. In practice, systems of three satellites are used to give visibility at all times. This kind of orbit is used for telephone relay satellites, at present mainly by the USSR, but could evidently also be employed for remote sensing.

Fig. 9.7. Ground track of a satellite in a geosynchronous orbit with non-zero inclination i. The figure-of-eight pattern is traced out once a day.

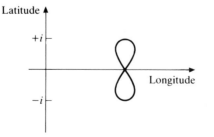

(c) Non-rotating orbits

For some purposes it may be desirable that the orbit's perigee remain at the same latitude, for example the maximum northerly latitude of a Molniya orbit. This can only be accomplished if $\omega = 0$ which, from (9.7), requires that $i = i_c$ where

$$\cos (i_c) = 5^{-\frac{1}{2}}$$

therefore

$$i_c = 63.4° \text{ or } 116.6°$$

(d) Sun-synchronous orbits

The earth itself is in orbit about the sun, anticlockwise when viewed from above the North Pole. Thus the sun appears, from the earth, to cross the field of the 'fixed stars' from west to east with a period of 1 year $(3.1558 \times 10^7 \text{ s})$. To put this another way, the sun appears to rotate about the earth with an angular speed $\Omega_s = 1.991 \times 10^{-7} \text{s}^{-1}$. If the satellite's precession $\Omega = \Omega_s$, its orbit will follow the sun. This kind of orbit is called sun-synchronous, and its practical implication is that the satellite will cross a given latitude at the same solar time every day. This is obviously useful for some kinds of environmental and scientific remote sensing satellites where it may be desirable to standardise light levels, amount of solar heating, and so on. In fact, sun-synchronous orbits have been very widely used for remote sensing satellites.

By substituting $\Omega = \Omega_s$ into (9.6) the semimajor axis a_{ss} of a sun-synchronous orbit can be shown to be given by

$$(a_{ss}/1000 \text{ km})^{7/2} = -6624.6 \cos i/(1 - e^2)^2 \qquad (9.9)$$

Thus if $a = 7878$ km (i.e. an orbital height of about 1500 km) and $e = 0$, i must be 102°. Note that sun-synchronous orbits are necessarily retrograde, and that they cannot have inclinations less than about 96° (otherwise a_{ss} would be less than the earth's radius).

The choice of crossing time for a sun-synchronous satellite using a visible-wavelength sensor depends on the type of observations to be made. A low sun-angle emphasises topographic effects, whereas a high sun angle (crossing time near noon) gives maximum illumination, which is useful for surfaces of low reflectance but may lead to saturation of the sensors over high-reflectance materials such as ice (Dowdeswell & McIntyre 1986). The crossing time itself varies, for a given (exactly-repeating sun-synchronous) orbit, with latitude. This can be calculated using equations (9.5), and the result is shown in fig. 9.8.

(e) Altimetric orbits

If a satellite is to be used for altimetry, it is desirable that the ascending and descending sub-satellite tracks should cross at about 90°. This is so that orthogonal components of the surface slope can be determined with equal accuracy. The crossing angle χ at the equator is given by the following expression, if the effects of the earth's departure from sphericity are ignored and the satellite's orbit is circular:

$$\cot(\chi/2) = |\cot i - 2\pi a^{3/2}(GM)^{-1/2}P_{\mathrm{E}}^{-1}\operatorname{cosec} i| \qquad (9.10)$$

Thus to obtain χ approximately 90°, an inclination of about 40° or 130° is needed. Polar or near-polar orbits give rather small crossing angles near the equator, and so are not particularly suitable for altimetry in that region, but the crossing angle increases towards the polar regions and as a fortunate consequence it is possible to select an orbit which not only has a high enough inclination to reach the polar areas but also has a reasonable crossing angle there (see fig. 9.9).

There is, or may be, a second criterion for altimetric orbits, especially for the study of the ocean surface. This relates to the choice of the repeat period, i.e. the temporal interval at which the altitude of a given point is measured. If this altitude is not subject to change then the problem does not arise, but the earth's ocean and, to a lesser extent, solid surfaces are subject to tidal action which causes them to vary in height. These

Fig. 9.8. Local time at which a sun-synchronous satellite crosses a given latitude, assuming a crossing time of midnight at the ascending node. The figure has been calculated for an orbital inclination of 99°, but will be similar for any low-altitude sun-synchronous orbit.

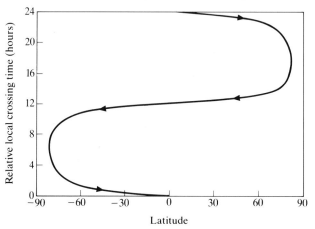

variations have dominant terms with periods of 12 and 24 hours, but there are also significant components with periods of one month and one year. The most important deep-water tides are, in order of diminishing amplitude, M_2 (the principal lunar tide, with a period of 12.421 hours), K_1 (the luni-solar tide, with a period of 23.934 hours), S_2 (the principal solar tide, with a period of 12.000 hours) and O_1 (the diurnal lunar tide, whose period is 25.819 hours).

Suppose we try to observe the ocean surface with an altimeter which repeats its track every 3.00 days. It will clearly measure the height at exactly the same point in the 12 and 24 hour cycles, and will thus entirely fail to measure these components, in a manner similar to the well known stroboscopic effect. Thus the profile obtained will be an unrepresentative, essentially instantaneous, picture of the topography.

Now let us suppose that we try to solve this problem by increasing the repeat period very slightly to 3.01 days. After one orbital repeat the 12 hour cycle will have occurred 6.02 times, and the 24 hour cycle 3.01 times,

Fig. 9.9. Angle at which sub-satellite tracks cross one another, as a function of latitude, for a satellite in a low earth orbit. The curves are labelled with the values of the orbital inclination.

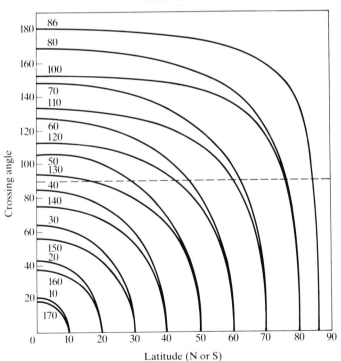

and we will be sampling points effectively only 0.02 and 0.01 of a cycle, respectively, away from the first measured point. It will thus take 50 orbital repeats, or about 150 days, to sample one complete 12 hour cycle, and twice as long to sample the 24 hour cycle. This phenomenon is called *aliasing* (from the Latin *alias*, at another time). We can state the result generally: If we try to measure a periodic phenomenon of frequency f_1 by making repeated measurements at frequency f_0, we will in fact observe the variation at a frequency f_a, the aliased frequency, given by

$$f_a = f_1 - f_0 \text{INT}(f_1/f_0 + 1/2) \tag{9.11}$$

In this expression, $\text{INT}(x)$ is the greatest-integer function, defined as the greatest integer not exceeding x. We note from this expression that the maximum frequency which can be present in the aliased spectrum is $f_0/2$, which is of course the Nyquist frequency (see section 2.3). If tides are significant in the investigation being planned, it is important that the orbit be chosen so that they are not aliased into inconvenient frequencies. Thus for example no orbit which repeats after an integral number of days will be suitable for the observation of S_2, since it will be aliased to a frequency $f_a = 0$. As another example, observations every three days (i.e. $F_0 = 0.013889 \text{ hr}^{-1}$) will alias O_1 into a frequency $f_a = -0.002935 \text{ hr}^{-1}$, so that observation for at least 14 days will be needed in order to sample one cycle.

(f) Exactly repeating orbits

We often require an orbit which exactly retraces its path over the earth's surface so that, for example, a given test-site or series of such sites may be revisited. For this to occur we need to arrange that

$$P_N(\Omega_e - \Omega) = 2\pi n_1/n_2 \tag{9.12}$$

where n_1 and n_2 are integers, having no common factor, representing respectively the number of days and the number of orbits between repeats. Ω_e is the earth's rotational speed $2\pi/P_E$.

For example, the reference orbit for the ERS-1 mission has the following orbital parameters: $a = 7153135$ m, $e = 0.00117$, $i = 98.5227°$. This is very close to being a sun-synchronous orbit (as substitution into (9.9) shows), so $\Omega = 1.997 \times 10^{-7} \text{s}^{-1}$, and substitution into (9.7) gives $P_N = 6027.9$ s. Thus the ratio n_1/n_2 has a value of 0.06977, whose reciprocal is $14\frac{1}{3}$. Hence $n_1 = 3$ and $n_2 = 43$, and the satellite revists a given point on the earth's surface every 3 days, after having made 43 orbits.

Because both P_N and Ω depend on a, i and e, fine tuning of the ratio on the right-hand side of (9.12) is possible, giving rise to a wide range of possible recurring orbits. Such orbits are often used for remote sensing satellites.

It will be apparent that, in the choice of repeat period for an exactly-repeating orbit (or even for a non-exactly repeating orbit), a balance must be struck between the requirements of frequent revisiting of a given site, and of a dense spatial coverage of ground tracks. That is to say, the spatial and temporal sampling frequencies are inversely related so that, for example, a 3-day repeat period for a low earth orbit gives ground tracks separated by about 900 km at the equator, whereas a 30-day repeat period will give tracks separated by about 90 km. Clearly in choosing the orbital repeat period for a given satellite mission, these considerations will need to be combined with a knowledge of the swath width of the sensor.

As an example, let us again consider the orbit of the ERS-1 satellite. The ratio n_1/n_2 (in (9.12)) is monotonically dependent upon the satellite's semimajor axis, and constraints imposed by the amount of fuel to be carried for the thrusters will probably limit the ratio to

$$11/158 \leq n_1/n_2 \leq 7/100$$

From the conditions associated with (9.12) we can see that (for example) the only exactly repeating orbit possible with a period n_1 of 3 days is that with $n_2 = 43$, whereas for $n_1 = 80$ we may accept $n_2 = 1143$, 1147 or 1149. For the case $n_1 = 3$, $n_2 = 43$, adjacent suborbital tracks are spaced at $360°/43 = 8.4°$ in longitude, corresponding to about 930 km at the equator and about 460 km at 60° north or south. For the case $n_1 = 80$ the spacing of adjacent suborbital tracks is 0.31° in longitude, corresponding to 35 km at the equator. Thus, a hypothetical sensor with a swath width of 40 km would give complete coverage of the earth's surface, within the satellite's latitudinal limits, for an 80-day repeat, whereas it would not do so with a three-day repeat.

To continue our example, we see that the choice of an 80-day repeat gives a high spatial density of sampling at the expense of a low temporal rate of one sample every 80 days. However, this limitation may not be as severe as it appears. If we consider the orbit with $n_2 = 1147$, we see that after three days the satellite has made 43.0125 orbits, which is (of course) very close to the 43 orbits it would have made if it had been placed in an exact 3-day orbit. Thus after 2.999128 days, the satellite has made exactly 43 orbits whereas the earth has rotated through 1079.69 degrees. The orbit is therefore only 0.31° east of its original position. Thus, in this particular case, adjacent suborbital tracks are traced out at three-day intervals, so that a wide-swath sensor will be able to keep a given point on the earth's surface in view for a number of consecutive overpasses. This type of orbit is often called a *drifting subcycle*, since what is strictly a long-period repeat may also be regarded as a slowly drifting orbit with a short repeat period. (Note that not all long-period repeat orbits have this property.) It is not, however, suitable for all applications: for example, the remarks made earlier on the subject of the aliasing of tidal frequencies still apply.

One final remark should be made before we leave the topic of the relationship between the spatial and temporal sampling yielded by a satellite orbit. Up to this point, we have assumed that the sensor carried by the satellite looks in a fixed direction relative to the satellite's velocity vector. If, however, the sensor is steerable (possibly by tilting the entire satellite), it may be possible to view the same region of the earth's surface during a number of orbital passes. This approach, adopted by the SPOT-1 satellite, clearly allows a temporary modification of the spatial and temporal sampling intervals, increasing the frequency of observation of a given region at the expense of neighbouring regions. The technique also has the advantage of providing data from different incidence angles, which may, as we have seen in earlier chapters, allow substantially more information to be extracted.

As a summary of the various types of orbit used in remote sensing, fig. 9.10 illustrates the principal ranges of semimajor axis, eccentricity and inclination.

9.3.5 *Decay of orbits and orbital lifetimes*

We have already mentioned the influence of atmospheric friction on launch vehicles, and clearly such friction will also act on an orbiting satellite. Ultimately, loss of energy through the action of atmospheric drag will cause the satellite to fall back to earth, or to burn up in the atmosphere, and this provides an upper limit to the useful lifetime of a satellite.

It can easily be shown that the reduction δa in the semimajor axis a of a satellite in a circular orbit, for each orbit about the earth, is given by

$$\delta a \approx 4\pi A \rho a^2 / M \qquad (9.13)$$

where A is the satellite's cross-sectional area (normal to its motion), ρ is the atmospheric density at the satellite's altitude, and M is the satellite's mass. The approximation in (9.13) is attributable to uncertainty in the

Fig. 9.10. Summary of the important satellite orbits. (*a*) Low earth orbits ($e \approx 0$). *a* is the semimajor axis, *h* the height above the earth's surface and P_N the nodal period. Some of the more important remote sensing satellites to use these orbits are named on the diagram. (*b*) Low and high earth orbits, including the effects of eccentricity but neglecting inclination.

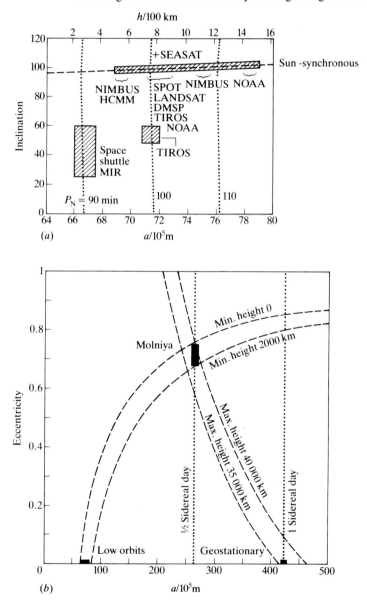

Fig. 9.11. Variation with height above the earth's surface of atmospheric pressure, density and temperature. (*a*) is plotted on a logarithmic scale to 50 000 km altitude, and (*b*) shows on linear scales the behaviour in the lowest 20 km of the atmosphere. In (*b*), the pressure and density are normalised to their standard sea-level values of 101.3 kPa and 1.225 kg m^{-3} respectively.

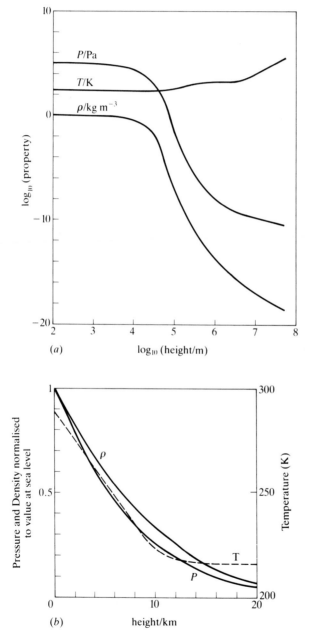

drag coefficient of a particular satellite, but the equation should be accurate in most cases to within a factor of about two.

As an example of the use of this equation, let us consider the LANDSAT 5 satellite. This has a mass of about 1700 kg, and a cross-sectional area of about 10 m². Its orbital height is about 700 km ($a = 7086$ km), at which height the atmospheric density, shown in fig. 9.11, is about 10^{-13} kg m^{-3}. Thus from (9.13), the satellite descends by about 0.4 m per orbit, or about 5 m per day. Observations of changes in the orbital parameters of satellites have been used extensively in this way for the determination of the properties of the earth's outermost atmosphere.

The maximum useful life of a satellite would be expected, on the basis of (9.13), to be proportional to M/A for a given orbital configuration. To a good approximation, we may write

$$\tau = Mf(h,e)/A \qquad (9.14)$$

where τ is the lifetime, and f (in units of m² s kg^{-1}) is a function of only the perigee altitude h and the eccentricity e. Fig. 9.12 summarises the behaviour of the function $f(h,e)$, although it should be remembered that the properties of the atmosphere at large distances from the earth are strongly affected by external influences such as the sun, and are in consequence variable.

Fig. 9.12. Approximate behaviour of the function $f(h,e)$ defined in the text.

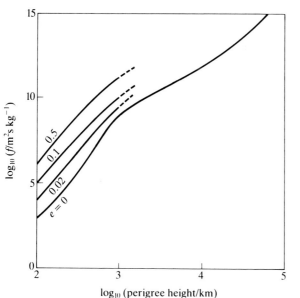

From fig. 9.12 and (9.14) it is clear that the maximum useful lifetime of LANDSAT 5 is expected to be about 500 years. Naturally, in this case, failure of the equipment carried by the satellite will be the limiting factor. However, if we consider a similar satellite in an orbit at an altitude of 100 km, we see that its expected lifetime would be only 2 days. Very low orbits (below about 150 km) are thus only used for military re-connaissance, where a spatial resolution of the order of 0.5 m is essential. The best spatial resolution achievable by such systems is not widely advertised, but it is probably 10 cm, or perhaps better, using large-diameter optics under good atmospheric conditions (see sections 2.4 and 3.4.3).

We should note also that, because the density of the earth's upper atmosphere varies with the effects of solar and geomagnetic activity, the prediction of satellite orbits is made difficult. For example, at an altitude of 800 km, an error of the order of 1 km in the along-track direction is likely to have built up after only two days (Agrotis 1988). At higher altitudes, solar radiation pressure is the dominant influence on satellite orbits, but again, prediction is hindered by its variability.

Problems

1. Write a short essay discussing the advantages and disadvantages of artificial satellites for remote sensing.
2. Consider a satellite in orbit about a spherically symmetric planet. If the orbit is circular, the subsatellite point travels uniformly along its circular track. If the orbit is not circular, however, the subsatellite track is still circular, but the subsatellite point moves along it at a variable rate. Show that the maximum along-track error in calculating the position of the subsatellite point on the assumption that it moves uniformly is about eD, where e is the eccentricity of the orbit and D is the planet's diameter.
3. For what semimajor axis would a satellite follow an orbit which is both sun-synchronous and non-rotating?
4. Derive a general expression for the direction of the subsatellite track, assuming a circular orbit. Show that at a latitude of 70° this direction is about 27° from north in the case of ERS-1, assuming that the satellite will have an inclination of 98.5° and a nodal period of 6028 s.
5. Estimate the lifetime of a cylindrical satellite of mass 2000 kg and diameter 5 m at an altitude of (a) 200 km, (b) 2000 km.

10

Data processing

10.1 Introduction

The general direction of this book has been that in which information travels, from the thermal or other mechanism for the generation of electromagnetic radiation, to its interaction with the surface to be sensed, thence to its interaction with the atmosphere, and finally to its detection by the sensor. It is clear that the data have not, however, reached their final destination. Firstly, they are still at the sensor and not with the data user. Secondly, the 'raw' data will in general require a significant amount of processing before they can be applied to the task for which they were acquired.

In this chapter we shall discuss the more important aspects of the processes to which the raw data are subjected. For the most part, it will be assumed that the data have been obtained from an imaging sensor, so that the spatial form of the data is significant. The principal processes are transmission and storage of the data, 'preprocessing', enhancement, and classification. The last three processes are generally regarded as aspects of the general subject of *image processing*. This, in particular, is a vast and well worked field of study, and we shall be able to do no more than outline its principal features. The reader who wishes to know the subject in greater detail is recommended to study the books by Swain & Davis (1978), Hord (1982), Schowengerdt (1983), Mather (1987) and Muller (1988), as well as the more general works of Curran (1985) and Harris (1987). The papers by Harris (1980) and by Townshend & Justice (1981) also provide a good introduction to the subject.

10.2 Storage and transmission of data

It is evident that the data must be brought from the sensor to the place where they are to be analysed. In the case of airborne remote sensing this is unlikely to present a problem, since missions are comparatively short and it is relatively easy to transport the data, whether they are recorded on photographic film or digitally on magnetic tape. Similar remarks apply to spaceborne observations from low-altitude reusable platforms such as the space shuttle, but when we consider a high-altitude satellite, which may be operational for five years or more and cannot generally be retrieved, the situation is obviously different. The data can either be transmitted to a terrestrial receiving station as they are detected, or they can be stored on-board, to be 'dumped' to a receiving station when convenient.

Each approach has its associated difficulties. A surface-based receiving station cannot receive signals from a satellite which is below the horizon, and usually requires that the satellite's elevation angle be greater than 5° in order to minimise atmospheric effects. This limits the angular distance ϕ, measured on the earth's surface, from the sub-satellite point to the receiving station. From fig. 10.1 it is clear that the satellite altitude h, its angular altitude θ, and the earth's radius R are related to ϕ by the following expression:

$$\cos(\theta + \phi) = R \cos \theta / (R + h) \tag{10.1}$$

Thus if $h = 800$ km and $\theta = 5°$, $\phi \approx 23°$. Such a receiving station will thus be able to receive data from a satellite 800 km high, up to about 2500 km

Fig. 10.1. Calculation of the relationship between the elevation angle θ of a satellite and the angle ϕ between the sub-satellite point and the receiving station.

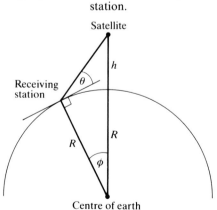

from the subsatellite point. Fig. 10.2 shows, by way of an example, the currently existing network of receiving stations for LANDSAT data, with their corresponding reception areas. Since the LANDSAT instruments are nadir-looking and there is no on-board storage of data, the reception areas define the areas of which images may be obtained.

One partial solution to the problem represented by the patchy coverage of fig. 10.2 is to employ a series of *relay satellites*, such as the TDRS (tracking and data relay satellite) system currently being deployed by NASA. These satellites, to which satellite data are sent, and from which they are then relayed to earth, are in geostationary orbit, thus giving coverage (by (10.1)) to about latitude $\pm 80°$, and a corresponding range of longitude at the equator. To date, two TDRS satellites have been deployed, one above longitude 41° W and the other above 171° W. The METEOSAT and GOES geostationary meteorological satellites are also used as relay satellites for data from buoys, land stations, balloons, aircraft etc. METEOSAT is located above longitude 0°, and the two GOES satellites above longitudes 135° W and 75° W.

Another solution, mentioned above, is the use of on-board data storage. In this way, up to one orbit's-worth of data can be stored and then transmitted to a suitable receiving station when the latter is in view by the satellite. The problem with this approach is usually that of

Fig. 10.2. Existing and planned coverage of the earth's surface by LANDSAT receiving stations (as at December 1988).

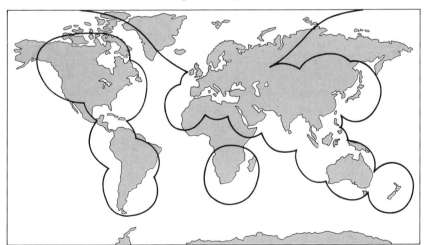

mechanical reliability of the tape recorders. The method was adopted for LANDSATs 1–3, but the recorders failed early during their respective missions.

Whichever of these methods is adopted, we must give some consideration to the rate at which data are recorded by a sensor, in order to calculate the transmission rate and the storage requirements. For a sensor of swath width w and resolution element of area A (we shall assume that the resolution elements cover the swath completely), carried by a platform whose *ground-track* speed is v, the mean rate at which resolution elements are viewed is vw/A. If n bits of data and k spectral bands (or equivalent, e.g. states of polarisation) are recorded from each resolution element, the *minimum* data rate is

$$DR_{min} = knvw/A \qquad\qquad (10.2)$$

Insertion of some typical values into (10.2) gives minimum data rates for satellite-borne sensors ranging from of the order of $10\,\mathrm{kb\,s^{-1}}$ ($1\,\mathrm{b} = 1$ bit) for a radar altimeter to $10\,\mathrm{Mb\,s^{-1}}$ for a SAR or high resolution visible imaging system. The actual data rates will be higher than those calculated by (10.2), by an amount dependent on the detailed instrumental design and degree of oversampling which is thought desirable. For example, the SAR system which will be carried by ERS-1 will generate data at a rate of $105\,\mathrm{Mb\,s^{-1}}$.

A typical computer magnetic tape can store about $0.5\,\mathrm{Gb}$. This corresponds to about 10 orbit's-worth of data at $10\,\mathrm{kb\,s^{-1}}$, or $30\,\mathrm{km}$ of data at $100\,\mathrm{Mb\,s^{-1}}$, for a satellite in a low orbit. This provides some idea of the volume needed for storing data in archives. Fast tape recorders with total capacities of about $5\,\mathrm{Gb}$ can be carried aboard satellites, and it is clear that such a storage device can hold one orbit's-worth of data from sensors whose combined data rate does not exceed about $1\,\mathrm{Mb\,s^{-1}}$. Such sensors may therefore be defined as *low data-rate sensors*, and their output can be stored for transmission, once per orbit, to a suitable receiving station. Note that for this to be possible, the system must be capable of transmitting its stored data in the comparatively short time during which a receiving station is in view (about 13 minutes if the angle ϕ in (10.1) is $23°$ and the orbital period is about 100 minutes).

10.3 Image processing

As was mentioned earlier, image processing is generally considered to consist of the three steps of preprocessing, image enhancement, and classification. Roughly speaking, these involve, respectively, the removal of systematic errors in the data, increasing their intelligibility (often by the removal of random errors) as a representation of the sensed object, and extracting meaningful patterns from the data. As will be apparent from this brief description, and more so from the more detailed descriptions which follow, the distinctions between these steps are not clear cut. This is one of the justifications for regarding image processing as a single subject whose purpose is to extract meaningful, preferably quantitative, patterns from the detected data. Another such justification is of course the very wide applicability of such techniques outside the field of remote sensing.

Most image processing is now carried out using digital data, since it is much easier to perform all but the simplest operations on data held in a computer memory, and because many remote sensing systems now generate data in digital form (see e.g. Ince, 1983). We shall for the present define an image to be a two-dimensional array of numbers, each of which represents the radiation intensity of one element of the surface. Each region of the image defined in this way is called a *picture element* or *pixel*, although the corresponding resolution element on the ground is some-times also called a pixel. (Some attempt has been made to introduce the term *rezel*, by analogy with pixel, but this does not seem to have found much support.) The numbers defining the radiation intensity are often referred to as *digital numbers* (*DN*s), since the data are in digital form, or *grey levels*, since a single band of data could be represented as a photographic image in which a given *DN* is transformed to a certain photographic density. (In fact, the term grey level is slightly misleading, for two reasons. The first is that a positive photograph will record a high intensity as white, and a low intensity as black, so that the *DN* is inversely related to the photographic density. The second reason why the term grey level is misleading is that much image processing is now performed on multispectral images, which are presented in colour. In this case, the colour of a displayed pixel is composed of red, green and blue components, the intensity of which is proportional to the *DN* of the corresponding spectral band.) In general, then, an image can be described as a matrix. If the image is monospectral, the matrix is two-dimensional and has m (rows) times n (columns) elements, each of which is the *DN* of the corresponding pixel. For a multispectral (or multipolarisation, or multitemporal) image with k channels (e.g. spectral bands), the matrix is three-dimensional, with m times n times k elements.

There is no intrinsic problem in modifying this definition to include continuous data such as photographic images, but it will be more convenient to make the above assumptions.

10.3.1 Preprocessing

As described at the beginning of this section, the purpose of preprocessing is to remove systematic errors from the data. The most important preprocessing operations are the correction of radiometric and geometric errors, i.e. calibration of the detected signal and registration of the image data with true surface positions. We should also include under this heading the initial stages of processing synthetic aperture radar data (chapter 8), which require substantial unscrambling before they form a spatially registered array of backscattering coefficients (i.e. a radar image). However, we will not add here to the remarks made in chapter 8 about SAR processing.

10.3.1.1 Radiometric correction

The data should in most cases be calibrated (although this is unusual for photographic images). Usually the type of calibration which is required is *radiometric* calibration, in which the relationship between the detector output and the input radiation intensity is established. For visible wavelength and radiometric systems, calibration is often performed automatically by the sensor, which periodically looks at a standard source such as the sun or an internal black body. (It should, however, be noted that such calibration will not permit correction to be made for atmospheric propagation effects such as the reduction in image contrast described in chapter 4.) Active microwave systems are often calibrated against a target of known backscattering cross-section, such as a corner-cube reflector (three adjacent sides of a cube, constructed from a reflecting material). Such correction may be necessary, for example to correct the output for a known non-linearity in the sensor calibration, or to allow for topographically induced variations in the intensity of the pixels. However, in many cases it will be sufficient if the sensor calibration is known to be monotonic and reasonably stable over time, so that data values may meaningfully be compared from one image to another.

10.3.1.2 Geometric correction

Geometric correction involves relating spatial coordinates on the image to the corresponding spatial coordinates on the surface, and can be achieved using data on the platform's position and direction of view, if these are available, at the time the image was recorded. For greater accuracy, however, ground control points (GCPs) are needed, as discussed briefly in chapter 9. The use of GCPs is particularly important if images are to be joined together to form a mosaic, or if images of the same area (acquired at different times or with different sensors) need to be compared with one another.

Geometric correction of an image using GCPs can either be carried out piecemeal, using a network of GCPs and interpolating the correction between them, or (which is usually adequate) it can be done by fitting a model to the entire image. If we write (x,y) for the row and column coordinates of a pixel in the image, and (x',y') for the coordinates (e.g. latitude and longitude, or grid coordinates) of the corresponding point on the ground, the model can be expressed generally as

$$x' = F(x,y)$$
$$y' = G(x,y)$$

The simplest useful model is one which assumes that the image is related to the object by a combination of rotation, a change of scale (different in the x and y directions) and shift of origin. These are all linear transformations, so the functions F and G may be expressed quite simply:

$$\left. \begin{array}{l} x' = a_1 + a_2 x + a_3 y \\ y' = a_4 + a_5 x + a_6 y \end{array} \right\} \tag{10.3}$$

In principle, three GCPs are sufficient to determine the model's parameters a_1 to a_6, although naturally a greater number is desirable as a precaution against random errors. A more general model allows for the possibility of oblique viewing and skew (which arises, for example, from the way in which a whiskbroom sensor operates – see chapter 5), which gives a variable scale across the image and introduces quadratic terms (in x^2, y^2 and xy) into equations (10.3). The minimum number of GCPs is then six, although in practice one would use as many as possible, and preferably of the order of 50 to 100, to correct geometrically a large image. Some kind of least-squares method is used to obtain the best fit to the data.

Once the transformation has been obtained, the image can be resampled onto a grid aligned with the desired ground axes x' and y'. This resampling is carried out by interpolation of the old pixels onto the new grid. Straightforward linear interpolation gives an unsatisfactorily disjointed image, and more time-consuming methods such as cubic convolution are often used. (See e.g. Curran, 1985.)

Geometric correction may also be necessary for atmospheric refraction, especially in the case of high resolution photogrammetric satellite observations. Because of the decrease with height of atmospheric density, rays are bent towards the earth's surface (fig. 10.3). For incidence angles less than about 45°, the following formula, derived from data quoted by Moccia & Vetrella (1986), is applicable to satellite observations (i.e. those which look through the whole atmosphere):

$$\Delta x \approx 2.5 \sin \theta \, (\cos \theta)^{-4.5} \text{ metres} \tag{10.4}$$

10.3.2 *Image enhancement*

Improvements to the image can be divided into two types: those which operate on individual pixels without reference to spatial context, and those which make use of spatial information. The first type can generally be referred to as *contrast modification*, and the second as *spatial filtering*.

10.3.2.1 *Contrast modification*

The contrast of a feature in an image (although it can also be defined for a whole image) is the ratio I_{max}/I_{min}, where I_{max} and I_{min} are the maximum and minimum *DN*s present in the feature. This is the sense in which contrast is to be understood when it is said that it is reduced by atmospheric scattering, especially at the blue end of the visible spectrum. Clearly such scattering will tend to redistribute the radiation from bright features into nearby directions, decreasing the maximum and increasing the minimum intensities. (Note that if we would rather not define contrast in such a way that it can become infinite, we may instead use the ratio $(I_{max} - I_{min})/(I_{max} + I_{min})$.) The contrast of 'raw' LANDSAT-type images is usually rather low in any case, because the instrument has to be

Fig. 10.3. Differential refraction in the atmosphere causes an apparent shift Δx in the position of an object.

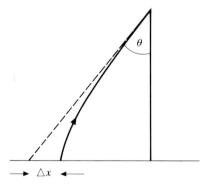
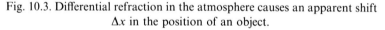

designed to accept a wide range of conditions of illumination. In consequence, the input irradiance which just results in saturation is set comparatively high, and most of the data will yield values concentrated in the lower half of the sensitivity range.

It is evident that the detectability of a feature to a human observer will be improved if the contrast is increased, although of course the information content of the image is unaltered by this process. The general object is to increase the use of the available display DN-space. Contrast stretching (or more generally contrast modification) is the process of transforming the DNs of all the pixels of an image according to some rule, replacing the 'image' DN by a 'display' DN (see fig. 10.4). The process can be represented as a graph of I_{dis} against I_{im}, that is, display DN against image DN, and it is clear that if this graph has a slope greater than 1, the contrast is increased. Note that an even greater use of display DN space is obtained if the data are first transformed into their principal components (see section 10.3.3.3 and Rothery, 1987) before contrast modifications are performed.

Let us assume for sake of argument that our image is stored on a computer whose memory and display both have a resolution of 8 bits. Thus both I_{im} and I_{dis} can be represented as integers from 0 to 255. It is clear that as we increase the contrast of an image, so we shall decrease the range of image DNs which will generate a useful (unsaturated) display DN. For example, if the transformation has a slope of 10, this range will be at most 25 units wide. Thus, caution must be exercised, and contrast modification is usually a task which requires some experience to achieve optimum results. Fig. 10.5 illustrates a typical linear contrast stretch, which is the simplest type of contrast modification.

A slightly more sophisticated transformation is the two-part linear stretch, but this may be regarded as a particularly simple kind of non-linear transformation. A more common non-linear transformation is that of *histogram equalisation*. To explain this technique we must introduce the concept of the *image histogram*, which is one of the simplest descriptors of the radiometric information contained in the image. It is simply a graph of the frequency distribution of the DNs present in the image, and clearly a histogram may also be defined for the display DNs. The process of histogram equalisation attempts to generate a histogram of display DNs which is uniform. It can easily be shown that the transformation function in this case is a suitably scaled version of the *cumulative* histogram of image DNs (see fig. 10.6).

Fig. 10.4. Image enhancement by contrast modification. The image shown in (*a*) is a 16-kilometre square extract from a single band of a LANDSAT scene of Portsmouth, England. In (*b*), the same data are shown after the application of a linear contrast stretch. The information content of the image is unchanged, but the intelligibility is improved. (Courtesy of National Remote Sensing Centre, UK.)

Fig. 10.5. A typical linear contrast stretch. The graph shows the transfer
function relating the displayed DN I_{dis} to the image DN I_{im}. In this particular
example, the contrast has been increased by a factor of 2.55.

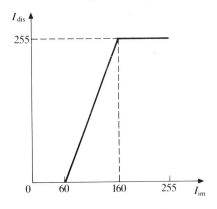

Fig. 10.6. Histogram equalisation. (a) is the image histogram, showing the
number of pixels with a given value of I_{im}. (b) is the cumulative histogram,
showing the total number of pixels (in thousands) with a DN below I_{im}. (c) is
the transfer function relating display DN to image DN. It is scaled copy of (b).
(d) is the resulting display histogram, showing the number of pixels with a
particular value of I_{dis}.

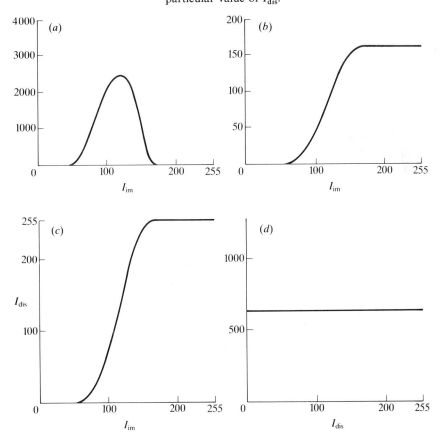

This transformation has the advantage that it increases the contrast to the greatest extent for the largest number of pixels, and it has no adjustable parameters so it can be applied simply and automatically. It is not invariably the most useful contrast modification, however, since it is undiscriminating. It may be the case that the part of the image in which one is most interested is a small fraction of the whole, for which a 'manual' contrast stretch (i.e. applied using an operator's skill rather than some automatic process) is more useful.

10.3.2.2 Suppression of random noise

Any image will be subject to random noise; that is, the *DN* of a given pixel will be subject to statistical uncertainty from a variety of causes (instrumental, atmospheric and so on). We have no way of guessing *a priori* what correction should be applied to the *DN* on this account, so we are forced, if we wish to reduce the noise level, to form some kind of average. This will be especially important in the case of coherent imaging systems, in which speckle is a significant contributor to image noise (see section 8.4.2).

A simple and fairly obvious technique of noise suppression is to replace the *DN* of a given pixel by the average of itself and its neighbours. We can represent this diagrammatically by a grid of boxes, each box representing a pixel, the central box representing the pixel to be processed, and the number in each box representing the weight of the contribution made by that pixel to the total sum (fig. 10.7). If the total of all the weights is unity, the overall brightness of the image will be unchanged.

Fig. 10.7 represents only one particular example of this type of averaging. We could for example include more pixels, and also 'taper' the

Fig. 10.7. Weighting distribution for a uniform 3×3 spatial average of an image.

$\frac{1}{9}$	$\frac{1}{9}$	$\frac{1}{9}$
$\frac{1}{9}$	$\frac{1}{9}$	$\frac{1}{9}$
$\frac{1}{9}$	$\frac{1}{9}$	$\frac{1}{9}$

weights away from the central pixel. It is clear, however, that the desired effect of suppressing noise will bring with it the usually undesirable effect of smoothing out the spatial structure of the image. For example, a sharp transition between two homogeneous regions will be spread by the average defined in fig. 10.7 over a width of three pixels (see fig. 10.8). Clearly the greater the reduction in noise, the larger this smoothing effect will be.

10.3.2.3 Spatial filtering

The box average discussed in the previous section is a simple example of spatial filtering. In general, such transformations can be regarded either as operations on the image, or operations on the Fourier transform of the image. Fig. 10.7, and all the cases we shall deal with in this section, involve *convolution* of the image with some function, and thus the Fourier transform of the image is multiplied by the Fourier transform of this function. Since the function is two-dimensional its Fourier transform will be likewise, and thus a function $a(\mathbf{q})$ of the vector spatial frequency $\mathbf{q}$. However, we can obtain a convenient one-dimensional representation of its nature by plotting the *average* value of $a(\mathbf{q})$ for all values of $\mathbf{q}$ with the same modulus q. This is done in fig. 10.9 for the function defined in fig. 10.7. Not surprisingly, fig. 10.9 resembles closely the sinc function defined in section 2.3. Since the spatial frequency spectrum (Fourier transform) of the image is multiplied by this function, we can see that the latter acts as a form of *low-pass filter*, suppressing the high spatial frequencies of the image.

It follows, then, that if we wish to detect edges in an image, that is, sharp transitions between uniform regions, we should reject low spatial frequencies and keep high frequencies. This is achieved using a *high-pass filter*, an example of which is shown in fig. 10.10 in both representations. As before, we may either think of the image as being convolved with the function represented by fig. 10.10(*a*), or of the Fourier transform of the image being multiplied by the function shown in fig. 10.10(*b*).

The filter shown in fig. 10.10 will remove areas of uniform intensity from the image, leaving only the sharp edges (see fig. 10.11). This is useful for the *detection* of sharp edges, but we might instead wish, as an aid to the interpretation of the image, to *enhance* them. The human eye responds mainly to edges, and finds particular difficulty in discriminating between adjacent uniform areas of very similar brightness unless they are separated by a boundary. To emphasise sharp edges in an image without

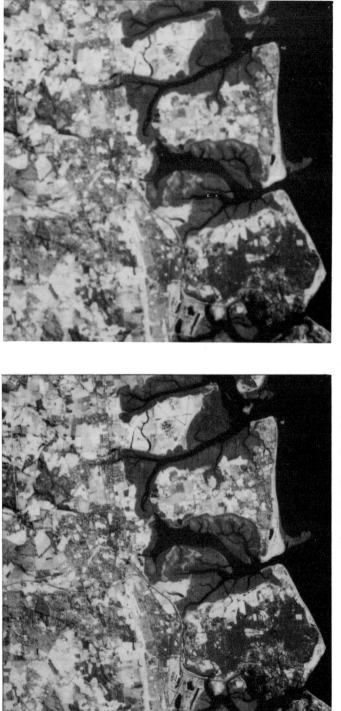

Fig. 10.8. Effect of applying a moving average to an image. The left-hand image shows a 16-kilometre square extract from a single band of a LANDSAT image of Portsmouth, England. The right-hand image shows the same extract after the application of the filter shown in fig. 10.7. The noise level is reduced, and the image is smoothed (blurred). (Reproduced by courtesy of the National Remote Sensing Centre, UK.)

changing uniform areas we need to apply a *high-boost filter*, which increases the higher spatial frequencies while leaving the lower ones unchanged. Fig. 10.12 illustrates such a filter, and fig. 10.13 demonstrates its effect.

Spatial filtering, as we saw in section 10.3.2.2, can be used to 'clean' an image by the removal of random noise. It may be used to remove *periodic noise*, such as the stripes often present in LANDSAT images because of the slightly different calibration of each of the sensors operating at a particular wavelength. Other examples of periodic noise are afforded by

Fig. 10.9. Representation of the function defined in fig. 10.7 as its Fourier transform. The Fourier transform of the image is multiplied by this function. Since the function diminishes in size with increasing spatial frequency q, it is known as a low-pass filter.

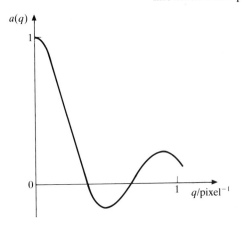

Fig. 10.10. A spatial high-pass filter, and its representation in the spatial frequency (Fourier transform) domain.

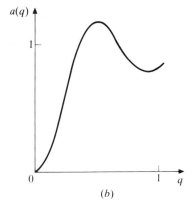

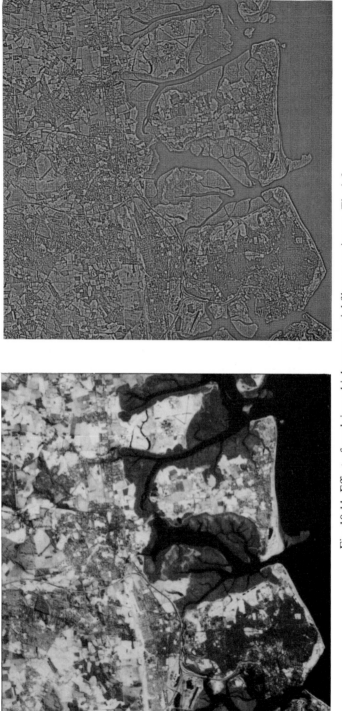

Fig. 10.11. Effect of applying a high-pass spatial filter to an image. The left-hand image shows a 16-kilometre square extract from a single band of a LANDSAT image of Portsmouth, England. The right-hand image shows the same extract after the application of the filter shown in fig. 10.10. Uniform regions have been removed, leaving sharp edges and fine-scale variations. (Reproduced by courtesy of the National Remote Sensing Centre, UK.)

radar images subject to time-periodic electromagnetic interference, and images which have been resampled from another system and contain raster lines. The unwanted features will appear prominently in the Fourier transform of the image, and may be removed in this domain. The 'cleaned' transform can then be retransformed to obtain an improved image.

Before we leave this brief discussion of spatial filtering procedures we should mention *gradient filters*. These respond to a local rate of change of *DN* with pixel position. In general, two such filters will be required, in order to extract the components of the *DN*-gradient in each of two perpendicular directions. The simplest pair of gradient filters is the Roberts filter, shown in fig. 10.14.

10.3.3 Image classification

Classification is the process whereby an image is converted into some kind of *thematic map*, in which regions with similar properties are indicated in the same way. At the simplest level this involves identifying pixels with similar *DN*s (which we assume implies that the corresponding areas of the surface are also similar to each other). It is thus an attempt to replace visual interpretation of an image by quantitatively based decisions. At a more advanced level, these pixels will be identified, grouped into continuous regions, recognised as features, textures and so on. In general, however, we can say that the classification of an image is an exercise in pattern recognition, whether or not the 'pattern' includes spatial information or just radiometric information.

Fig. 10.12. A spatial high-boost filter, and its representation in the spatial frequency (Fourier transform) domain.

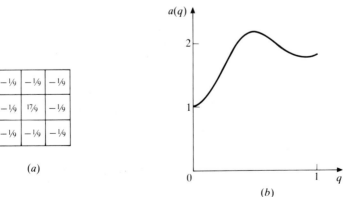

(a)

(b)

Fig. 10.13. Effect of applying a high-boost spatial filter to an image (edge enhancement). The left-hand image shows a 16-kilometre square extract from a single band of a LANDSAT image of Portsmouth, England. The right-hand image shows the same extract after the application of the filter shown in fig. 10.12. Uniform regions are unchanged, but sharp edges and fine-scale variations are emphasised. (Reproduced by courtesy of the National Remote Sensing Centre, UK.)

10.3.3.1 Density slicing

Density slicing is the simplest technique for the classification of images. The image is simply divided into regions in which the *DN*s lie between certain limits. It is thus essentially a monospectral technique, and is of limited usefulness in most cases. It is most often combined with spatial filtering such as high-pass or gradient filtering, when a simple threshold is all that is required (for example, to identify regions of the image in which the detected edges are significantly above the level of random noise). It is also useful in dealing with, for example, thermal infrared observations of the sea surface, where the image intensity has a well defined dependence on a single physical parameter (in this case the SST). Other applications of density slicing include bathymetry (see chapter 5), and the generation of masking datasets before further image analysis proceeds. (For example, one might wish to apply a contrast stretch to only those parts of an image representing water surfaces. A simple density slice would allow those pixels corresponding to land surface to be identified and ignored.) Plate 9 shows an example of density slicing. In general, however, classification makes use of multispectral data.

10.3.3.2 Multispectral classification

We saw in chapter 3 that many materials have a characteristic *spectral signature*, and this is the basis of multispectral classification. If every possible surface material had a unique, known and invariable spectral signature, and if we could construct a noiseless sensor which could measure this spectrum over all wavelengths, correcting the data for incidence and scattering angle effects, atmospheric attenuation and scattering, there would be no fundamental problem in identifying by remote sensing the contents of any resolution element, assuming it to be homogeneous (that is, that that resolution element does not contain more than one material). Multispectral classification attempts to solve the

Fig. 10.14. A pair of Roberts filters, used for extracting the gradient of an image.

0	−1
1	0

−1	0
0	1

problems caused by the failure of real systems and materials to meet these conditions. Plate 10 shows an example of the end result of such a multispectral classification process.

We shall assume for simplicity that only two spectral bands have been recorded, although in practice the number will usually be greater than this. If the *DN* recorded for a given pixel in spectral band 1 is I_1, and that recorded in band 2 is I_2, we can plot these values on a graph of I_2 against I_1. If we repeat this for all the pixels of an image, we hope that the data will form a series of *clusters*, such as in fig. 10.15.

Our modification of the 'ideal' situation described in the preceding paragraph is to assume that each cluster represents a unique surface type. We are thus left with the problems of finding the clusters, and of identifying the surface types to which they correspond. It is important to remark at this point that the extent to which the data will fall into distinct clusters is dependent on the spatial resolution of the system. The main reason for this is that, as pixels increase in size, the noise level diminishes (see Townshend & Justice 1981).

The simplest type of multispectral classification is to form a *band ratio*, that is, I_2/I_1. This is often done to effect a crude removal of viewing angle effects and *vignetting* (a decrease in image brightness away from its centre, caused by the geometry of the optical system). Because of the poor behaviour of the ratio I_2/I_1 when $I_1 \ll I_2$, a *transformed band ratio* is sometimes used. The commonest transformation is to take the arctangent of the ratio I_2/I_1. Once the band ratios have been formed, they can be examined by density slicing. However, it is clear that this simple technique does not make full use of the clustering which (we hope) exists in the data.

Fig. 10.15. Clusters of points in *DN*-space. Each point represents one pixel, and I_1 and I_2 are the *DN*s of the pixel in each of two spectral bands. The clusters are identified with classes of data.

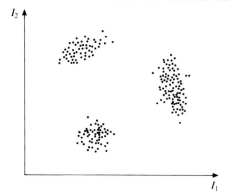

Unsupervised classification is an automated process for finding clusters of data points in *DN*-space, possibly using only a subset of the whole image, and then deciding to which cluster any pixel belongs. Various standard methods are available for finding clusters, and one important method (the *K-means algorithm*) will be described below. The result of an unsupervised classification will be the identification of sets of 'similar' pixels, but of course the method will be unable to say whether one set of pixels represents one surface type or another. In fact, it would be more correct to call these unsupervised techniques data *allocation* techniques. A more fruitful approach is *supervised classification*, in which the operator defines homogeneous *training areas*, known (by ground-truth fieldwork) to contain the surface material of interest. The cluster in *DN*-space corresponding to this surface material is thus defined, in a sense, by the operator. Curran and Williamson (1985) have discussed the quantitative data requirements for this approach, suggesting that the number of ground-resolution elements necessary to define a class should be at least 30, and that the area which should be sampled at each such element should be $D^2(1 + 2G)^2$, where D is the linear size of the ground-resolution element and G is the accuracy of the geometric correction of the image.

Once the clusters have been defined, a method must be chosen for determining to which cluster a given pixel belongs. There are essentially three methods in use. The simplest is the *minimum-distance algorithm*. In this case, the distance of a given point in *DN*-space is calculated from the centroid of each cluster, and the minimum distance is used to identify the required cluster. This is illustrated schematically in fig. 10.16. The method has obvious disadvantages, notably that no account is taken of the shape of the clusters, and that the 'distance' between two points in *DN*-space is generally not a quantitatively significant measure. To counter this specific objection, other definitions of distance in *DN*-space have been developed.

The *box classifier* attempts to take account of these defects. A 'box', that is, a closed figure, is drawn round each cluster, and a pixel is ascribed to the corresponding class if it falls within the box (fig. 10.17). In our simple two-dimensional case the box will be a plane figure. The commonest shape is a rectangle with sides parallel to the axes, but parallelepipeds or more complicated figures may be used.

The boxes defined by the training data may overlap, in which case an unambiguous classification cannot be made for some pixels. They will also probably not fill all available *DN*-space, so that some pixels may not be classified. Both of these facts are likely to be realistic reflexions of the amount of information yielded by the training areas.

A substantial refinement of the box classifier is provided by the last method which we will describe, the *maximum-likelihood classifier*. In this case, the probability that a pixel belongs to a given class is calculated, and the pixel is assigned to whichever class is more likely. If the probability falls below some preset level for all classes, the pixel may be rejected as unclassifiable, although this is not an invariable rule. The probability is assumed to be proportional to the density of training points in *DN*-space, so it is especially important that the training areas should be large enough. This method is usually the most accurate, and clearly the most expensive to use.

It was mentioned earlier that the commonest method of defining clusters in an unsupervised classification is the *K*-means algorithm (also known as the isodata method). In this approach, the user specifies the number of clusters to be identified, and initial guesses (although these may be generated automatically) as to where, in *DN*-space, the centroids of these clusters are located. The algorithm then calculates to which cluster each pixel belongs, on the basis of which centroid it is nearest to. The centroids of these new clusters are calculated, and the process is repeated. It is thus an iterative method (as are most unsupervised classifiers), and must be supplied with a criterion of acceptable convergence otherwise it will probably go round its loop for ever.

Fig. 10.16. Minimum-distance classifier. A point in *DN*-space is assigned to a class on the basis of which of the class centroids *A, B, C* or *D* lies nearest to it. This means in practice that the class boundaries are as shown by the dashed lines. The class centroids *A, B, C* and *D* are identified by the use of training areas, or by an unsupervised classification.

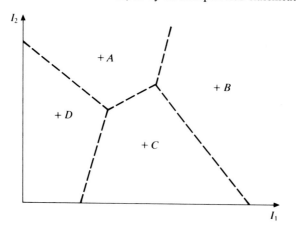

The choice of the number of clusters to be sought in a classification is a matter for intelligent guesswork. The accuracy of the classification process will decrease with the number of clusters, but on the other hand the potential usefulness of the classification will increase. Thus there will be some optimum number of clusters. Mathematical treatments of this aspect of the problem have been developed, involving the use of 'separability statistics' (see Swain & Davis, 1978).

It is perhaps useful finally to give a crude characterisation of supervised and unsupervised classification techniques. A supervised technique which uses field data will, by definition, give useful and meaningful classes. The clusters in *DN*-space may overlap, in which case the classes will not be identifiable from the data. Unsupervised techniques, on the other hand, will by definition yield identifiable classes in the data. These may, however, be difficult or impossible to relate to real features on the earth's surface.

10.3.3.3 Principal and canonical components

A multi-band (e.g. multispectral, multitemporal, multipolarisation, etc.) image will in many important cases show a significant degree of correlation between bands. This means that to display all the bands of data is in some sense wasteful, and if, as is generally the case, the display unit of the image processing system is capable of representing only three bands (via the three primary colours) and the image contains more than

Fig. 10.17. The box classifier makes greater use of information about the shape of the clusters.

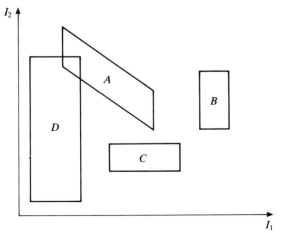

three bands, this correlation will be distinctly unhelpful. Such correlation can have a wide variety of causes. If the image is multispectral, the bands may overlap, spatial variations in atmospheric propagation or viewing geometry may affect all of them in similar ways, or there may be a real physical correlation between the measured properties. If the image is multitemporal, and correctly registered (see section 10.3.1.2), little physical change may have occurred in most resolution elements between the times when the data were gathered.

One approach to this problem is to *transform* the image data by producing new bands which are linear combinations of the old ones. Two main types of transformation are used – the *principal components transformation* (PCT – otherwise known as the *Karhunen–Loeve transformation*) and the *canonical components transformation* (CCT). These are chosen, respectively, to remove the correlation between bands, and to maximise the separability of previously defined classes of data. They may thus form preprocessing steps before classification of an image. In order to explain how these transformations work, we must first define a suitable notation for them.

Let us suppose that we have an image consisting of N pixels, for each of which B bands of data have been recorded, and that $I_i(p)$ represents the value of the ith band for the pth pixel. The *covariance matrix* c_{ij} of the image is defined as

$$c_{ij} = \frac{1}{N} \sum_{p=1}^{N} (I_i(p) - \langle I_i \rangle)(I_j(p) - \langle I_j \rangle) \tag{10.5}$$

where

$$\langle I_i \rangle = \frac{1}{N} \sum_{p=1}^{N} I_i(p) \tag{10.6}$$

is the mean value in the ith band. The covariance matrix is thus a two-dimensional array with B^2 terms.

By inspection of (10.5), we can recognise that the diagonal terms c_{ii} are the variances σ_i^2 of the bands, and the off-diagonal terms $c_{ij}(i \neq j)$ are related to the correlation coefficients ρ_{ij} between the ith and jth bands by

$$c_{ij} = \rho_{ij} \sigma_i \sigma_j \tag{10.7}$$

The transformation operations consist of replacing the band value $I_i(p)$ for the pth pixel by a new value $I_i'(p)$ which is a weighted sum of all the old values $I_j(p)(j = 1 \ldots B)$ for that pixel. This can be represented as a matrix operation:

$$I_i'(p) = \sum_{j=1}^{B} M_{ij} I_j(p) \tag{10.8}$$

Again, M_{ij} (the transformation matrix) is a two-dimensional array with B^2 terms.

The object of the PCT is to remove entirely the correlation between bands, i.e. to diagonalise the covariance matrix. By convention, the new bands I_i' are designated so that I_1' has the largest variance, I_2' the next largest, and so on:

$$c_{11}' > c_{22}' > \ldots > c_{BB}' \tag{10.9}$$

This process is easiest to visualise for a two-band image ($B=2$), where the $I_i(p)$ can be plotted as a two-dimensional scatter diagram in the manner of fig. 10.15. The PCT can then be seen to be a simple rotation of the coordinates in *DN*-space, as shown in fig. 10.18.

It is generally found with LANDSAT images of land surfaces that the first principal component (I_1') contains a high proportion, usually about 90%, of the total variance. Thus this single transformed band contains most of the information in the original multi-band image. The first three principal components together often contain over 98% of the total variance, so that very little information is sacrificed by displaying only these three transformed bands. In particular, much of the image *noise* is contained in the lowest-order components, which can be used to remove it. Plate 11 shows the principal components of an image, and plate 12 shows these components combined to form a false-colour image.

Fig. 10.18. The Principal Components Transformation. I_1 and I_2 are the *DN*s of the original (untransformed) image in each of two spectral bands. I_1' and I_2' are the transformed *DN*s, such that there is the least correlation between them.

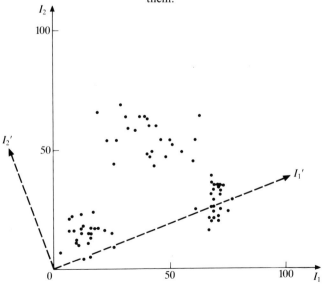

It is apparent (e.g. from fig. 10.18) that the PCT takes no account of the clustering of the data which one hopes to be present. The CCT is similar to the PCT, but it relies on the existence of a number of previously defined clusters (classes – see section 10.3.3.2), and in effect performs a PCT on them. In this case, the covariance matrices are calculated for each class separately and then averaged to give a mean covariance matrix $\langle c_{ij} \rangle$. The between-class covariance matrix b_{ij} is also calculated, using the class means (centroids). The desired transformation M_{ij} is then that which simultaneously diagonalises the matrix b_{ij} while reducing $\langle c_{ij} \rangle$ to the identity matrix. It thus attempts to produce clusters of points in DN-space which are (in two dimensions) circular, and maximally separated from each other in a direction parallel to the I_1' axis.

In practice, objections can be raised to both of these transformations. The combination of different bands implied by (10.8) is in general a step away from, rather than towards, the ideal of data which directly represent physical quantities. The CCT in addition relies on (a) a realistic separation of the original data into classes, and (b) the assumption that the clusters representing these classes in untransformed DN-space are of similar shape and size. Despite these limitations, however, both PCT and (to a lesser extent) CCT represent powerful methods of analysing multiband data.

10.3.3.4 *Texture analysis*

Although multispectral analysis has received the greatest attention as a method of classification, it makes no use of the spatial information present in the image. In general, methods which incorporate this information can be classed as texture analysis. This often aids human photointerpreters, and can sometimes be used in machine classification. Texture is difficult to define precisely (it is the 'frequency of change and arrangement of the tones in an image'), but one obvious although inadequate measure is the local variance of the image DN, and another is the shape of the spatial frequency spectrum. Such techniques have so far largely found application in the analysis of radar imagery, especially SAR, where multispectral data are usually unavilable. At the simplest level, the texture of an image (visible or radar) may contain information about the surface geometry. Foody (1986) and Rees & Dowdeswell (1988) have given examples of this.

Any numerical definition of true texture, suitable for automatic extraction methods, will be some measure of coarseness in the image. This eliminates the simplest possible definition, which is the local variance of the *DN* measured in a 'window' centred on the pixel of interest. Such a definition clearly contains no information about spatial frequencies. However, it is possible to define measures of texture which operate directly on the *DN*s of the pixels within such a window, for example by calculating the local autocorrelation function or by constructing a *semi-variogram*. The latter is measured along a linear transect through the data, and is defined as

$$\frac{1}{2M} \sum_{p=a}^{a+M-1} [I(p) - I(p+h)]^2$$

M is the number of pixels in the transect, and *p* enumerates them, beginning with pixel *a*. The semi-variogram is then a plot of this quantity as a function of the separation distance *h*. It has the advantages that it is easily calculated, and reasonably easily interpreted. (See Curran (1987), Ramstein & Raffy (1989) for recent discussions of the application of the semi-variogram to remote sensing.)

More sophisticated measures of texture have also been developed, of which the most important is probably the *spatial dependence matrix* (also called the co-occurrence matrix or the transition matrix). This is a square matrix with n^2 elements, where *n* is the number of values which a *DN* can take (e.g. 256 for 8-bit data, which obviously raises substantial practical difficulties by virtue of the enormous amount of calculation needed to generate it). A matrix element a_{ij} denotes the frequency with which a *DN* of *i* occurs in the adjacent pixel to a *DN* of *j*. Thus, for example, a perfectly homogeneous region will give rise to a spatial dependence matrix which is diagonal. From this matrix, various scalar measures of texture can be defined.

It is naturally true that potentially the most powerful methods of classifying images will make use of both multispectral and textural analysis. Indeed it is clear that a typical noisy 'raw' image, after classification, will be likely to yield an improbably high number of spatially homogeneous regions which have been assigned to a single class, and some form of texture analysis is needed to reduce this number. One such approach is the merging of adjacent regions which satisfy some criterion of similarity.

There is one further aspect to texture which we should mention before we leave the topic. This is the use of *syntactic methods* for recognising features in digital imagery. This is an attempt to reproduce the way a

human interpreter recognises objects, by giving the computer a set of rules
for the identification of objects of a certain type. These rules follow the
form '*if* there is a pixel at (x_1,y_1) with the property A *and* there is a pixel at
(x_2,y_2) with the property B, *then* the pixels form part of an object C. Such
rules are extremely difficult to put into practice.

10.4 Data compression

A digital image can easily consist of several Mb of data or more, and by its
very size therefore presents difficulties in storage and transmission.
Methods by which this quantity of information can be reduced are
therefore of particular interest. The subject of data compression might
more logically have been presented within section 10.2, but it is more
convenient to discuss it here, following on from the discussion of
classification. The reason for this is that the classification process itself
(and indeed the process of principal components transformation) may be
regarded as a form of data compression, since fewer bits of data are
required to specify a classified image than the raw data from which it is
derived. In general, methods of data compression can be divided into
reversible processes from which all the original information can be
retrieved, and *irreversible* processes of which this is not true. The example
given above, of classification, is clearly an irreversible process.

There are two main types of reversible data compression. The first is
known as *Huffman coding*, and makes no use of spatial information in the
image (see Hord 1982). Instead, it changes the representation of the *DN*s
in the image in such a way that the most frequently occurring *DN*s are
assigned binary numbers containing only a few bits, whereas the
infrequently occurring values are assigned longer binary numbers.
Clearly, since different *DN*s are represented by binary numbers of
different lengths, care must be exercised that there is no possibility of
ambiguity in the way a string of bits is decomposed into its constituent
parts, and the Huffman code ensures this lack of ambiguity. In this way,
the total number of bits needed to specify the image is reduced, by an
amount depending on the *entropy* of the original image. The entropy is
defined to be

$$\frac{-\Sigma p_i \log_e p_i}{\log_e 2}$$

bits per pixel, where p_i is the frequency with which *DN* i occurs in the
image, and the summation is carried out over all possible values of i.

The other type of reversible compression employs spatial information in the image. It is evident that if an image consists of a 1000 by 1000 array of pixels, each of which has the same value of *DN*, Huffman coding can only reduce the data requirement to 1 bit per pixel, or 10^6 bits. On the other hand, the information content of the image is very much less than this, and *run-length encoding* can reduce the number of bits required. In essence, this consists of specifying a *DN*, and then the number of places within the data stream (the length of the run) for which this value is maintained. In the very unrealistic example which we have just given, the number of bits required to describe the image would be only a few dozen (20 bits to describe the run length of 1 000 000 pixels, plus a few more to specify the actual *DN* of the pixels and the 1000 by 1000 format of the image).

A more sophisticated version of run-length encoding, based on the idea of *tesseral addressing*, specifies the area and location of square regions within the image. Such a method is in principle capable of even greater savings of data, but it will be apparent that neither the one-dimensional nor the two-dimensional technique will effect much of a saving, if any, in the case of a noisy image, since the spatial changes of *DN* will be very frequent. Thus in the case of real images, the greatest savings of data are likely to be made using run-length encoding only in the case of classified images.

Irreversible processes of data compression are usually easier to implement. It is clear that sampling or averaging an image has the potential to reduce the data content, and in the case of sampling the mechanism is obvious. If we use only every other pixel in both the *x* and the *y* dimensions of the image, we will reduce its size, and the number of bits required to specify it, by a factor of 4. The corresponding loss is in the equivalent ground resolution of the image.

The sampling method which we have just described is in fact very wasteful, since it throws away some of the original data (three quarters of them in the example just given). If instead we were to calculate simple 2 by 2 box averages from the original image we should use these data, while retaining the factor of 4 reduction in the data content. Thus in general, smoothing (averaging) achieves the result we desire, although if the smoothed data are placed in an array of the same dimensions as the original image it is not clear how we should take advantage of the reduced data content, since, other things being equal, any array of a given size will require the same number of bits of data to define it.

One approach to this problem is to store the image as its Fourier transform. Normally, the Fourier transform will contain exactly the same amount of information as the image, and require the same number of bits to specify. However, if we merely *truncate* the transform, and possibly also apply a taper to it to reduce 'ringing' effects, we shall simultaneously reduce the data requirement and smooth the image which will be obtained when the transform is retransformed (see chapter 2). In fact the Fourier transform, although powerful when employed in this way, is not ideal, and use of the *Hadamard transform* is preferable for digital data.

Problems

1. A satellite is in a circular polar orbit about the earth, with an orbital radius of 7400 km. It transmits to earth a data signal with a carrier frequency of 5000 MHz. Calculate the minimum bandwidth of the receiver if this signal is to be detected at all times when the satellite is above the horizon.

2. A SAR is carried aboard a satellite in a low earth orbit. It has an imaging swath of 100 km, and a pixel size of 30 m square. If 10 bits of data are recorded per pixel, what is the minimum data rate?

3. Approximately how much information can be stored on a single 35 mm black and white photographic negative? Express this also in bits per cubic metre, and compare your figure with other storage devices (e.g. magnetic tape, video tape, floppy disc, compact disc, human brain?).

4. Show that the minimum number of relay satellites necessary to give complete coverage of the equator is three. How many satellites would be needed to give complete coverage between the Arctic and Antarctic circles? (Assume that the line of sight to the satellite must be at least $5°$ above the horizon.)

5. An image contains three ground control points. These have been placed at known coordinates on a grid reference system, namely (100, 400), (200, 100) and (300, 0) (all in metres with respect to the origin of a particular map). The image coordinates of these GCPs are (0.811, 0.868), (0.721, 0.568) and (0.746, 0.430) metres respectively. By assuming that the relationship between the ground and the image coordinates is linear, establish the grid coordinate corresponding to the pixel at (0.605, 0.618).

6. An image histogram is zero for DNs up to 50, rises uniformly to a region of constant value between DNs of 75 and 125, then falls uniformly to zero at DN 150. Calculate the values of transformed DN corresponding to DNs of 25, 50, 75, 100, 125 and 150 for (a) a simple linear stretch, and (b) histogram equalisation, if maximum use is made (without saturation) of a display unit which can display DNs up to 255.

7. Show that for a two-band image with pixel values $I_1(p)$ and $I_2(p)$ for the pth pixel, the principal components transformation is

$$I_1'(p) = I_1(p)\cos\theta - I_2(p)\sin\theta$$
$$I_2'(p) = I_1(p)\sin\theta + I_2(p)\cos\theta$$

where

$$\tan 2\theta = 2(\langle I_1\rangle\langle I_2\rangle - \langle I_1 I_2\rangle)/(\langle I_1^2\rangle - \langle I_1\rangle^2 + \langle I_2\rangle^2 - \langle I_2^2\rangle).$$

8. A (small) image consists of 64 pixels arranged in an 8×8 grid. In each pixel the brightnesses I_1 and I_2 in two spectral bands have been measured, with the following results:

BAND 1

50	39	34	42	55	85	77	75
51	31	26	20	46	66	77	61
67	18	22	15	20	40	71	75
28	26	24	24	34	62	80	70
67	74	47	24	29	28	72	83
64	81	75	34	35	32	56	39
78	66	70	62	55	25	60	61
77	73	53	32	60	50	47	45

BAND 2

52	50	37	42	55	30	37	25
55	40	69	80	45	70	33	65
61	72	72	78	82	45	32	69
37	80	75	75	80	68	34	65
32	41	50	78	71	78	30	23
37	27	30	45	72	65	60	50
40	42	30	65	51	64	63	58
45	35	64	45	55	58	50	55

(a) Construct histograms for I_1 and I_2. Do the pixels appear on this basis to be divisible into a number of classes?

(b) Perform a cluster analysis on the data. How many classes are represented in the image, and how many distinct regions does it contain? Use both manual calculation and the K-means algorithm.

(c) Find the principal components of the data.

(d) Find the entropy of each band.

APPENDIX

Data tables

This section collects together some of the more commonly needed data in remote sensing. Some of these data have appeared in the text; it is nevertheless felt that it will serve a useful purpose if they are all presented in one place.

Physical constants

c (speed of light *in vacuo*) by definition	$2.997\,924\,580 \times 10^8\,\text{ms}^{-1}$
h (Planck constant)	$6.626\,076 \times 10^{-34}\,\text{Js}$
e (charge on the electron)	$1.602\,177 \times 10^{-19}\,\text{C}$
m_e (mass of the electron)	$9.109\,390 \times 10^{-31}\,\text{kg}$
u (atomic mass unit)	$1.660\,540 \times 10^{-27}\,\text{kg}$
m_p (mass of the proton)	$1.007\,276\,\text{u}$
m_n (mass of the neutron)	$1.008\,665\,\text{u}$
μ_0 (permeability of free space) $= 4\pi \times 10^{-7}\,\text{Hm}^{-1}$	$1.256\,637 \times 10^{-6}\,\text{Hm}^{-1}$
ε_0 (permittivity of free space)	$8.854\,188 \times 10^{-12}\,\text{Fm}^{-1}$
Z_0 (impedance of free space)	$376.730\,\Omega$
G (universal gravitational constant)	$6.672\,6 \times 10^{-11}\,\text{N}\,\text{m}^2\,\text{kg}^{-2}$
k (Boltzmann constant)	$1.380\,658 \times 10^{-23}\,\text{J}\,\text{K}^{-1}$
c_W (Wien's constant)	$2.897\,755 \times 10^{-3}\,\text{K}\,\text{m}$
σ (Stefan–Boltzmann constant)	$5.670\,51 \times 10^{-8}\,\text{W}\,\text{m}^{-2}\,\text{K}^{-4}$

Units

The following table lists some of the units, together with their S.I. equivalents, which are in common use in remote sensing.

1 ångström	$10^{-10}\,\text{m}$
1 inch	$25.4\,\text{mm}$
1 foot	$0.3048\,\text{m}$
1 nautical mile	$1.852\,\text{km}$
1 hectare	$10^4\,\text{m}^2$
1 acre	$4.05 \times 10^3\,\text{m}^2$

1 pound	0.454 kg
1 knot	0.514 m s^{-1}
1 bar	100 kPa
1 atmosphere	101.3 kPa
1 torr = 1 mm Hg	0.133 kPa
1 electron volt	1.602 × 10^{-19} J
1 calorie	4.187 J
1 gauss	100 μT
1 gamma	1 nT

Photometric units

Photographic quantities are often measured in units weighted according to the notional wavelength sensitivity of the human eye, which is most sensitive to light to wavelength 0.55 μm and has a half-power bandwidth of approximately ±0.05 μm. This means that there is no direct conversion from photometric to radiometric quantities, but an approximate conversion within the visible waveband can be made:

Luminous flux	1 lumen (lm) ≈ 1.5 mW at 0.55 μm wavelength	
Luminous intensity	1 candela = 1 lm sr^{-1}	
Luminance	1 nit = 1 lm sr^{-1} m^{-2}	1 stilb = 10^4 nit
Illuminance	1 lux = 1 lm m^{-2}	1 phot = 10^4 lux

Typical values of the illuminance of a horizontal surface at the earth's surface, in lux, are as follows:

Noon, clear day	100 000
overcast	10 000
sunrise/set	500
sun 5° below horizon	5
10°	0.05
15°	0.003
20°	0.0004
Full moon, clear	0.5
cloudy	0.05
Starlit sky, clear	0.005
overcast	0.000 05

Properties of the sun

Radius	6.96 × 10^8 m
Mass	1.99 × 10^{30} kg
Total radiated power	3.85 × 10^{26} W
Black-body temperature	5770 K

Properties of the earth

Equatorial radius	6 378 135 m
Polar radius	6 356 775 m
Semimajor axis of orbit about sun	1.4960 × 10^{11} m
Mass	5.976 × 10^{24} kg
Angular velocity about axis	7.292 116 × 10^{-5} s^{-1}
GM	3.986 004 34 × 10^{14} m^3 s^{-2}
g (standard gravitational acceleration at surface)	9.806 65 ms^{-2}

Dynamic form factor (J_2)	$1.082\,63 \times 10^{-3}$
Mean albedo	≈ 0.35
Land area	$1.49 \times 10^{14}\,\text{m}^2$
Ocean area	$3.61 \times 10^{14}\,\text{m}^2$
Mean land elevation	$860\,\text{m}$
Mean ocean depth	$3.9\,\text{km}$
Tropical year (equinox to equinox)	$31\,556\,926\,\text{s}$
Sidereal year (fixed star to fixed star)	$31\,558\,150\,\text{s}$
Sidereal day	$86\,164.09\,\text{s}$

Mean properties of the earth's atmosphere

Density (sea level, 15 °C)	$1.225\,\text{kg}\,\text{m}^{-3}$
Pressure (sea level, 15 °C)	$1.013\,25 \times 10^5\,\text{Nm}^{-2}$
Thermal conductivity	$2.534 \times 10^{-2}\,\text{Wm}^{-1}\,\text{K}^{-1}$
Dry composition (major constituents) by volume	
Nitrogen	0.7809
Oxygen	0.2095
Argon	0.0093
Carbon dioxide	0.0003
Optical depth (sea level)	$0.008\,79\,(\lambda/\mu\text{m})^{-4.09}$
(valid for $0.01\,\mu\text{m} < \lambda < 0.4\,\mu\text{m}$)	
Refractive index (0 °C, sea level)	
visible	1.000 292

[At standard temperature and pressure (STP: 298 K and 101 325 Pa), and with a partial pressure of 1500 Pa of water vapour, the refractive index of air between 0.2 and 2 μm is given approximately by the expression $10^7(n-1) = 2720 + 14.689\lambda^{-2} + 0.238\lambda^{-4}$, where λ is the (free-space) wavelength in μm. For different temperatures and pressure, $(n-1)$ is proportional to p/T. The effect of changing water vapour content near STP is to reduce the refractive index by about 4×10^{-10} for an increase of 1 Pa in the partial vapour pressure. (Kaye & Laby, 1973)]

    radio   1.000 288

[The refractive index of air at radio frequencies is given approximately by the expression $10^6(n-1) = 103.49p_1/T + 177.4p_2/T + 86.26p_3/T(1 + 5748/T)$, where p_1, p_2 and p_3 are the partial pressures of dry air, CO_2 and H_2O respectively, all expressed in mm Hg. (Kaye & Laby, 1973)]

Mean annual sea-level temperature

Equator	27 °C
30° N or S	20 °C
60°	-2 °C
90°	-25 °C

Mean density of the ionosphere

The following table gives the typical average mid-latitude daytime density of free electrons in m^{-3}

D-layer	50 km	10^7
	90 km	10^{10}
E-layer		
	140 km	10^{11}
F_1-layer		
	250 km	10^{12}
F_2-layer		
	600 km	10^{11}

A typical value of the *total electron content* (TEC), obtained by integrating the number density with respect to height, is $3 \times 10^{17}\,\mathrm{m}^{-2}$, although like all of the above values this is subject to large variations (see e.g. Boyd, 1974).

Vapour pressure of water

The following table gives the vapour pressure p of water, in units of kPa, as a function of temperature T in °C.

T	p
0	0.6
5	0.9
10	1.2
15	1.7
20	2.3
25	3.2
30	4.2
35	5.6
40	7.4
45	9.6
50	12.3

Solar position

Since much environmental remote sensing is dependent on solar illumination, it is useful to be able to calculate the sun's position in the sky at a given date, time and position on the earth's surface. This will permit, for example, hours of darkness and solar elevation angles to be determined. Note that in all of the following expressions, it is assumed that angles are expressed in *degrees*.

Let Φ be the latitude (measured north of the equator), and Λ the longitude (east of the Greenwich meridian) or the site on the earth's surface. T is the time, measured in hours and expressed as Greenwich Mean Time, and d is the day number (January $1 = 1$, February $1 = 32$ etc.). Note that this is often erroneously called the *Julian Day* number. In fact, the Julian Day number is the number of days which have elapsed since noon on the first of January, 4713 BC.

The *equation of time* expresses the difference in position between the true sun and a fictitious mean sun which appears to move uniformly across the sky. It is given approximately by

$$E = 2.47 \sin\{1.97(d-80)\} - 1.92 \sin\{0.986(d-3)\} \tag{A1}$$

The sun's declination is given approximately by

$$\delta = \sin^{-1}\{0.3987 \sin(0.986[d-80])\} \tag{A2}$$

and its hour angle by

$$H = 15T - 180 + \Lambda + E \tag{A3}$$

Equations (A1) to (A3) give the sun's position as a function of time. The hour angle H and declination δ are converted to altitude a and azimuth A using equations (A4) and (A5).

$$a = \sin^{-1}\{\sin\delta\sin\Phi + \cos\delta\cos\Phi\cos H\} \qquad (A4)$$
$$A = \cos^{-1}\{(\sin\delta - \sin\Phi\sin a)/(\cos\Phi\cos a)\} \qquad (A5)$$

Note that there is a trigonometrical ambiguity involved in taking the inverse cosine in (A5). It is resolved by noting that A and H must have the same sign.

If we wish to calculate the times of sunrise and sunset, we must first find the sun's declination using (A2), and then use (A6) to find the sun's hour angles at rising and setting. These values can then be substituted into (A3) to find the times, accurate to a few minutes.

$$H_{rs} = \cos^{-1}\{(-0.014 - \sin\delta\sin\Phi)/(\cos\delta\cos\Phi)\} \qquad (A6)$$

BIBLIOGRAPHY

Agrotis, L.G. (1988). Near-Earth satellite orbit determination and applications. *ESA Journal*, **12**, 441.

American Society of Photogrammetry (1981). *Manual of Photogrammetry*, 4th edition. Falls Church, Virginia: American Society of Photogrammetry.

Anderson, J.M. & Wilson, S.B. (1984). The physical basis of current infrared remote-sensing techniques and the interpretation of data from aerial surveys. *International Journal of Remote Sensing*, **5**, 1.

Apel, J.R. (1983). A survey of some recent scientific results from the SEASAT altimeter. In *Satellite Microwave Remote Sensing*, ed. T.D. Allan, p. 321. Chichester: Ellis Horwood.

Barrett, E.C. & Curtis, L.F. (1982). *Environmental Remote Sensing*, 2nd edition. London: Chapman and Hall.

Baskasov, A.I. *et al.* (1984). Simultaneous radiometric and radar altimetric measurements of sea microwave signatures. *IEEE Journal of Oceanic Engineering*, **OE9**, 325.

Beckmann, P. & Spizzichino, A. (1963). *The Scattering of Electromagnetic Waves from Rough Surfaces*. Oxford: Pergamon Press.

Bleaney, B.I. & Bleaney, B. (1976). *Electricity and Magnetism*, 3rd edn. Oxford: Oxford University Press.

Bondi, H. (1988). Remote sensing: moving towards the 21st century. In *Proceedings of the 1988 International Geoscience and Remote Sensing Symposium*, p. 5. ESA SP-284. Paris: European Space Agency.

Boyd, R.L.F. (1974). *Space Physics*. Oxford: Clarendon Press.

Bracewell, R.N. (1978). *The Fourier Transform and its Applications*, 2nd edn. New York: McGraw Hill.

Brown, G.S. (1977). The average impulse response of a rough surface and its applications. *IEEE Transactions on Antennas and Propagation*, **AP25**, 67.

Carsey, F.D. (1985). Summer Arctic sea ice character from satellite microwave data. *Journal of Geophysical Research*, **90**, 5015.

Carslaw, H.S. & Jaeger, J.C. (1959). *Conduction of Heat in Solids*, 2nd edn. Oxford: Oxford University Press.

Chen, H.S. (1985). *Space Remote Sensing Systems*. Orlando, Florida: Academic Press.

Cheney, R.E. & Marsh, J.G. (1981). Oceanographic evaluation of geoid surface in the Western North Atlantic. In *Oceanography from Space*, ed. J.F.R. Gower, p. 855. New York: Plenum Press.

Chetty, P.R.K. (1988). *Satellite Technology and its Applications*. Blue Ridge Summit, Pennsylvania: Tab Books.

Christodoulidis, D.C. *et al.* (1985). Observing tectonic plate motions and deformations from satellite laser ranging. *Journal of Geophysical Research*, **90**, 9249.

Cihlar, J., Brown, R.J. & Guindon, B. (1986). Microwave remote sensing of agricultural crops in Canada. *International Journal of Remote Sensing*, **7**, 195.

Colwell, R.N. (1983). *Manual of Remote Sensing*, 2nd edn. Falls Church, Virginia: American Society of Photogrammetry.

Cracknell, A.P. (1981). *Remote Sensing in Meterology, Oceanography and Hydrology*. Chichester: Ellis Horwood.

Crane, R.G. & Anderson, M.R. (1984). Satellite discrimination of snow/cloud surfaces. *International Journal of Remote Sensing*, **5**, 213.

Curran, P.J. (1985). *Principles of Remote Sensing*. London: Longman.

Curran, P.J. (1987). What is a semi-variogram? In *Proceedings of the 13th Annual Conference of the Remote Sensing Society*, p. 36. Nottingham: Remote Sensing Society.

Curran, P.J. & Williamson, H.D. (1985). The accuracy of ground data used in remote-sensing investigations. *International Journal of Remote Sensing*, **6**, 1637.

Dainty, J.C. & Newman, D. (1986). Speckle from pseudo-random structures. In *Wave Propagation and Scattering*, ed. B.J. Uscinski, p. 281. Oxford: Clarendon Press.

Dowdeswell, J.A. & McIntyre, N.F. (1986). The saturation of LANDSAT MSS detectors over large ice masses. *International Journal of Remote Sensing*, **7**, 151.

Drewry, D.J. (1983). *Antarctica: Glaciological and Geophysical Folio*. Cambridge: Scott Polar Research Institute.

Elachi, C. (1987). *Introduction to the Physics and Techniques of Remote Sensing*. New York: Wiley-Interscience.

Engel, C.E. (1968). *Photography for the Scientist*. London: Academic Press.

Erich, U. & Gottschalk, D.H. (1984). ERS-1 and its potential for industrial utilization. *Earth-Oriented Applications of Space Technology*, **4**, 211.

Foody, G.M. (1986). Viewing geometry effects on SAR image tone and its importance for land cover mapping. In *Proceedings of a symposium held by Commission IV of the International Society for Photogrammetry and Remote Sensing and the Remote Sensing Society*, p. 240. Nottingham: Remote Sensing Society.

Forshaw, M.R.B. *et al.* (1983). Spatial resolution of remotely sensed images. *International Journal of Remote Sensing*, **4**, 497.

Foster, J.L. (1983). Night-time observations of snow using visible imagery. *International Journal of Remote Sensing*, **4**, 785.

Fraysse, G. (1984). Perspectives of remote sensing in Europe at the end of the decade. *Earth-Oriented applications of Space Technology*, **4**, 199.

Goetz, A.F.H. (1979). *Preliminary stereosat mission description.* Jet Propulsion Laboratory report no. 720-33, Pasadena, California: Jet Propulsion Laboratory.

Goetz, A.F.H. *et al.* (1985). Imaging spectrometry for Earth remote sensing. *Science*, **228**, 1147.

Gordon, H.R. & Morel, A.Y. (1981). Water colour measurements – an introduction. In *Oceanography from Space*, ed. J.F.R. Gower, p. 207. New York: Plenum Press.

Gower, J.F.R. (1981). *Oceanography from Space.* New York: Plenum Press.

Grassl, H. & Koepke, P. (1981). Corrections for atmospheric attenuation and surface reflectivity in satellite-borne SST measurements. In *Oceanography from Space*, ed. J.F.R. Gower, p. 97. New York: Plenum Press.

Guzkowska, M.A.J. *et al.* (1988). Satellite radar altimetry over arid regions. In *Proceedings of the 1988 International Geoscience and Remote Sensing Symposium*, p. 659. ESA SP-284. Paris: European Space Agency.

Hallikainen, M.T. (1984). Retrieval of snow water equivalent from NIMBUS-7 SMMR data: Effect of land-cover categories and weather conditions. *IEEE Journal of Oceanic Engineering*, **OE9**, 372.

Harris, R. (1980). Spectral and spatial image processing for remote sensing. *International Journal of Remote Sensing*, **1**, 361.

Harris, R. (1987). *Satellite Remote Sensing.* London: Routledge & Kegan Paul.

Hecht, E. (1987). *Optics*, 2nd edn. Reading, Massachusetts: Addison-Wesley.

Hord, R.M. (1982). *Digital Image Processing of Remotely Sensed Data.* New York: Academic Press.

Ince, F. (1983). Digital image processing systems and remote sensing. *International Journal of Remote Sensing*, **4**, 129.

Jackson, J.D. (1975). *Classical Electrodynamics*, 2nd edn. New York: Wiley.

Jepsky, J. (1985). Airborne laser profiling and mapping. *Lasers and Applications*, **4**, 95.

Justice, C.O. *et al.* (1985). Analysis of the phenology of global vegetation using meteorological satellite data. *International Journal of Remote Sensing*, **6**, 1271.

Kaye, G.W.C. & Laby, T.H. (1973). *Tables of Physical and Chemical Constants*, 14th edn. London: Longman.

Leese, O.A., Novak, C.S. & Clarke, B. (1971). An automated technique for obtaining cloud motion from geosynchronous satellite data using cross correlation. *Journal of Applied Meteorology*, **10**, 110.

Lipson, S.G. & Lipson, H. (1981). *Optical Physics*, 2nd edn. Cambridge: Cambridge University Press.

Long, M.W. (1983). *Radar Reflectivity of Land and Sea.* Dedham, Maryland: Artech House.

Longair, M.S. (1984). *Theoretical Concepts in Physics.* Cambridge: Cambridge University Press.

Marsh, J.G. *et al.* (1986). Global mean sea surface based upon the SEASAT altimeter data. *Journal of Geophysical Research*, **91**, 3501.

Massey, H. (1964). *Space Physics.* Cambridge: Cambridge University Press.

Mather, P.M. (1987). *Computer Processing of Remotely-sensed Images – an Introduction.* Chichester: Wiley.

Measures, R.M. (1984). *Laser Remote Sensing.* New York: Wiley.

Moccia, A. & Vetrella, S. (1986). An integrated approach to geometric prescision processing of spaceborne high-resolution sensors. *International Journal of Remote Sensing*, **7**, 349.

Muller, J.P. (1988). *Digital image processing in remote sensing.* London: Taylor and Francis.

Newton, D.C. (1989). Letter. *New Scientist*, 14 January, p. 72.

Omar, M.A. (1975). *Elementary Solid State Physics.* Reading, Massachusetts: Addison-Wesley.

Ouchi, K. (1986). On the imaging of ocean waves by synthetic aperture radar. In *Wave Propagation and Scattering*, ed. B.J. Uscinski, p. 297. Oxford: Clarendon Press.

Open University (1978). *Images and Information*, 2nd edn. Milton Keynes: Open University Press.

Parkinson, C.L. *et al.* (1987). *Arctic Sea Ice, 1973–1976.* NASA SP-489. Washington: National Aeronautics and Space Administration Scientific and Technical branch.

Ramstein, G. & Raffy, M. (1989). Analysis of the structure of radiometric remotely-sensed images. *International Journal of Remote Sensing*, **10**, 1049.

Raney, R.K. (1982). Processing synthetic aperture radar data. *International Journal of Remote Sensing*, **3**, 243.

Rao, U.R. & Chandrashekar, S. (1986). An international regime for remote sensing – problems and prospects. *International Journal of Remote Sensing*, **7**, 3.

Ray, R.E. & Fischer, W.A. (1957). Geology from the air. *Science*, **126**, 725.

Rees, W.G. (1988). Foreword. In *Radio Echo-Sounding as a Glaciological Technique*, comp. A.D. Macqueen. Cambridge: World Data Centre 'C' for Glaciology.

Rees, W.G. & Dowdeswell, J.A. (1988). Topographic effects on light scattering from snow. In *Proceedings of the IGARSS '88 Symposium, Edinburgh, September 1988.* ESP SP-284. Paris: European Space Agency.

Robinson, I.S. (1985). *Methods of Satellite Oceanography.* Chichester: Ellis Horwood.

Robinson, I.S., Wells, N.C. & Charnock, H. (1984). The sea surface thermal boundary layer and its relevance to the measurement of sea surface temperatures by airborne and spaceborne radiometers. *International Journal of Remote Sensing*, **5**, 19.

Rothery, D.A. (1987). Decorrelation stretching and related techniques as an aid to image interpretation in geology. In *Proceedings of the 13th Annual*

Conference of the Remote Sensing Society, p. 194. Nottingham: Remote Sensing Society.

Schanda, E. (1986). *Physical Fundamentals of Remote Sensing*. Berlin: Springer-Verlag.

Schowengerdt, R.A. (1983). *Techniques for Image Processing and Classification in Remote Sensing*. New York: Academic Press.

Sieber, A. & Noack, W. (1986). Results of an airborne synthetic-aperture radar (SAR) experiment over a SIR-B (Shuttle Imaging Radar) test site in Germany. *ESA Journal*, **10**, 291.

Slater, P.N. (1980). *Remote Sensing: Options and Optical Systems*. Reading, Massachusetts: Addison-Wesley.

Smith, D.E. *et al.* (1985). A global geodetic reference frame from LAGEOS ranging (SL5.1AP). *Journal of Geophysical Research*, **90**, 9221.

Srokosz, M.A. (1986). On the joint distribution of surface elevation and slopes for a nonlinear random sea, with an application to radar altimetry. *Journal of Geophysical Research*, **91**, 995.

Stewart, R.H. (1985). *Methods of Satellite Oceanography*. Berkeley, California: University of California Press.

Swain, P.N. & Davis, S.M. (1978). *Remote Sensing: The Quantitative Approach*. New York: McGraw Hill.

Tabbagh, A. (1973). Essai sur les conditions d'application des mesures thermiques à la prospection archéologique. *Annales de Géophysique*, **29**, 2.

Thomas, I.L. & Minnett, P.J. (1986). An introductory review of the measurement of ocean surface wind vectors with a satellite radar scatterometer. *International Journal of Remote Sensing*, **7**, 309.

Townshend, J. & Justice, C. (1981). Information extraction from remotely sensed data. *International Journal of Remote Sensing*, **2**, 313.

Trevett, J.W. (1986). *Imaging Radar for Resources Surveys*. London: Chapman and Hall.

Tsang, L., Kong, J.A. & Shin, R.T. (1985). *Theory of Microwave Remote Sensing*. New York: Wiley.

Ulaby, F.T., Moore, R.K. & Fung, A.K. (1981, 1982 and 1986). *Microwave Remote Sensing*, 3 vols. Reading, Massachusetts: Addison-Wesley.

Uscinski, B.J. (1986). *Wave Propagation and Scattering*. Oxford: Clarendon Press.

Valenzuela, G.R. (1978). Theories for the interaction of electromagnetic and oceanic waves – a review. *Boundary-layer Meteorology*, **13**, 61.

Van de Hulst, H.C. (1981). *Light Scattering by Small Particles*. New York: Dover publications.

Veck, N.J. (1985). Atmospheric transmission and natural illumination (visible to microwave regions). *GEC Journal of Research*, **3**, 209.

Wagner, C.A. (1985). Radial variation of a satellite orbit due to gravitational errors: implications for satellite altimetry. *Journal of Geophysical Research*, **90**, 3027.

Wahl, T., Eldhuset, K. & Aksnes, K. (1986). SAR detection of ships and ship wakes. In *SAR Applications Workshop*, p. 61. ESA SP-264. Paris: European Space Agency.

Warren, D. & Turner, J. (1988). Cloud track winds from polar orbiting satellites. In *Proceedings of the 1988 International Geoscience and Remote Sensing Symposium*, p. 549. ESA SP-284. Paris: European Space Agency.

Wooding, M.G. (1988). *Imaging Radar Applications in Europe: Illustrated Experimental Results 1978–1987*. ESA TM-01. Noordwijk: European Space Agency.

Yost, E.F. & Wenderoth, S. (1967). Multispectral colour aerial photography. *Photogrammetric Engineering*, **33**, 1020.

ANSWERS TO NUMERICAL PROBLEMS

2.1	$0.72\,\text{kV}\,\text{m}^{-1}$; $2.4\,\mu\text{T}$.
2.2	$0.03\,\text{m}$; $0.8\,\text{m}$.
2.3	$4.5\,\text{km}\,\text{s}^{-1}$.
2.4	$40\,\text{km}$; $3\,\text{km}$.
2.5 (a)	8.9; 0.025; 0.75; 0.49.
(b)	8.0; 2.2; 0.74; 0.48.
2.6	0.0500; 0.0463; 1.5×10^{-20}; 1.4×10^{-198}.
2.8	0.11%; $1100\,\text{K}$.
3.3	$2.66 \times 10^{17}\,\text{m}^{-2}$.
3.4	$2.82 \times 10^{-15}\,\text{m}$.
4.2	$(131.3, 119.0)$; $32.6\,\text{m}$; $37.1\,\text{m}$.
4.3	$885\,\text{m}$.
5.2	$0.040\,\text{K}^{-1}$; $0.015\,\text{K}^{-1}$.
5.3	Approx. February 4.
5.4	$34\,\text{K}$; $10\,\text{K}$; 0.5.
6.1	$10^{-8}\,\text{W}$.
6.2	30%; 38%.
6.5	$2\,\text{GHz}$.
8.1	10^{18}.
8.2	$10\,\text{m}\,\text{s}^{-1}$ at $55.6°$ or $124.5°$ from north (both with $180°$ ambiguity).
8.3	Approx. 1000.
8.4	$1.5\,\text{m}^2$; $14\,\text{m}^2$; $390\,\text{m}$; $1900\,\text{m}$ (ignoring earth's curvature).
9.3	$9815\,\text{km}$.
9.5	1 month; $30\,000$ years.
10.1	$0.24\,\text{MHz}$.
10.2	$8\,\text{Mb}\,\text{s}^{-1}$.
10.3	$10\,\text{Mb}$; $10^{14}\,\text{b}\,\text{m}^{-3}$. Typical values of $\log_{10}$ (data storage in $\text{b}\,\text{m}^{-3}$): book, 10. CCT and floppy disc, 12. Compact disc, 13. Video tape, 14. Human brain, 16.

10.4 4.

10.5 (77, 75).

10.6 (a) 0; 0; 64; 128; 191; 255.

(b) 0; 0; 43; 128; 213; 255.

10.8 3 classes; centroids at (50.6, 54), (25.1, 74.4), (74.7, 33.5). Principal components $0.80 \, I_1 - 0.60 \, I_2$; $0.60 \, I_1 + 0.80 \, I_2$. Entropies 5.27 bits per pixel, 2.62 bits per pixel.

INDEX

absorption length, 14, 57, 60
absorption lines, molecular, 65, 66–8,
    123
across-track direction, 156
active systems, definition of, 5, 131
advanced very high resolution
    radiometer, 108
aerial photography, history of, 3
aeroplane, invention of, 4
aerosols, 62, 66, 147
agriculture, applications of remote
    sensing to, 2, 4, 77, 104–5, 107, 154
aircraft, 4, 171–4, 194
albedo, 45–7, 147, 227
alias, 30, 134, 143, 184–6
along-track direction, 155
altimetric orbit, 184–6
ambiguity, range, 134, 163–4
amplitude, of electromagnetic wave, 10
amplitude transmission function, 30
anaglyph, 21, 90
analysis of features in aerial
    photographs, 81
angular frequency, 9
anomalies, geological, 2
anomaly, temperature, 113
antenna, 116
antenna temperature, 117
anthocyanins, 55
apogee, 176
Apollo 6, 4
archaeology, applications of remote
    sensing to, 92, 113, 145
area, effective, of antenna, 119

Aristotle, 3
ASA number, of photographic film, 77
ascending node, of satellite orbit, 177,
    178
atmosphere
    attenuation by, 62–8, 108, 123–4, 133,
        193
    composition of, 61–2, 227
    friction in, 175, 188–92
    luminance and illuminance, 92, 226
    pressure, 2, 62, 112, 190, 227
    properties of, 190, 227
    refraction in, 141, 199–200, 227
    sounding of, 2, 69–70, 129, 131, 135,
        147
    temperature of, 2, 62, 112, 129, 190,
        227
    turbulence in, 70–2
attenuation, 60, 63
    coefficient, 64
    length, 60, 63, 125–6, 145
avalanches, 2
AVHRR, *see* advanced very high
    resolution radiometer
azimuth direction, 155
azimuth shift, in SAR images, 164–5

backscattering cross-section, 149–50
Baird, J.L., 4
band gap of semiconductor, 103, 105
band ratio, 212
bandwidth, 30, 117, 121
barium sulphate, 48
baseline, in stereophotography, 88

238